Foundation Mathematics for Science and Engineering Students

Philip Prewett

Foundation Mathematics for Science and Engineering Students

 Springer

Philip Prewett [ID]
Department of Mechanical Engineering
University of Birmingham
Birmingham, UK

ISBN 978-3-030-91965-8   ISBN 978-3-030-91963-4   (eBook)
https://doi.org/10.1007/978-3-030-91963-4

This Springer imprint is published by the registered company Springer Nature Switzerland AG
The registered company address is: Gewerbestrasse 11, 6330 Cham, Switzerland

Preface

This compact book is intended to provide students of STEM subjects entering university with the foundation level understanding of mathematics they will need for their studies. Based upon an introductory mathematics for engineering module delivered by the author, it will refresh and extend what first year students must know and seeks to unify the knowledge and approach to the subject of students from different countries and educational backgrounds. In this way it will support students to make a successful transition from school to university—a change which many find challenging. Much of the content will also be useful for students still at school, studying for A-levels and similar international qualifications.

The book explains foundation mathematics from first principles, rather than as a set of given rules. The emphasis is upon promoting understanding and learning through examples why a particular method is applied. It was written during the United Kingdom's "lockdown" due to the COVID-19 pandemic when schools, colleges and universities were closed for an extended period and a generation of students was consequently disadvantaged. The book will be particularly useful to them, filling gaps in essential mathematical knowledge and understanding.

Philip Prewett
Dorchester-on-Thames, Oxfordshire, UK
March 2022

Acknowledgements

I am grateful to Mrs. Margaret Sloper who typed the book and laid out the text and equations so expertly.

I am indebted to my former students and graduate assistants on my Engineering Mathematics modules at Birmingham University and to my colleagues in the Department of Mechanical Engineering.

Contents

Trigonometry

Contents

© The Author(s), under exclusive license to Springer Nature
Switzerland AG 2022
P. Prewett, *Foundation Mathematics for Science and Engineering Students*,
https://doi.org/10.1007/978-3-030-91963-4_1

1

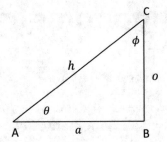

Fig. 1.1 Right angled triangle with opposite adjacent and hypotenuse labelled for angle θ

Referring to the right angled triangle in ▶Fig. 1.1, the Circular Functions are:

$$
\begin{aligned}
\sin\theta &= \tfrac{o}{h} & \operatorname{cosec}\theta &= \tfrac{h}{o} \\
\cos\theta &= \tfrac{a}{h} & \sec\theta &= \tfrac{h}{a} \\
\tan\theta &= \tfrac{o}{a} & \cot\theta &= \tfrac{a}{o}
\end{aligned}
$$

1.1 Trigonometric Relationships

▶Figure 1.2 is a useful schematic for trigonometric interrelationships. Angles are measured by anticlockwise rotation as shown. A few sample angles are shown. The signs of the adjacent and hypotenuse for each angle follow the standard Cartesian convention: positive above Ox and to the right of Oy, applied in each of the four quadrants. Using ▶Fig. 1.2, it is easy to see that

$$
\begin{aligned}
\sin\theta &= \cos(\tfrac{\pi}{2}-\theta) & \tan(\tfrac{\pi}{2}-\theta) &= \cot\theta \\
\cos\theta &= \sin(\tfrac{\pi}{2}-\theta) \\
\sin(\pi-\theta) &= \sin\theta & \tan(\pi-\theta) &= -\tan\theta \\
\cos(\pi-\theta) &= -\cos\theta \\
\sin(\pi+\theta) &= -\sin\theta \\
\sin(-\theta) &= \sin(2\pi-\theta) \\
&= -\sin\theta
\end{aligned}
$$

(The hypotenuse is always positive, regardless of quadrant.)

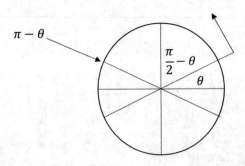

Fig. 1.2 Schematic tool for trigonometric relationships

1.2 Pythagoras' Theorem

Pythagoras' Theorem applies to all right angled triangles. It links the opposite, adjacent and hypotenuse relative to either of the non right angles.

$$h^2 = o^2 + a^2$$

This applies to both θ and ϕ in ►Fig. 1.1 with o and a interchanged.

Pythagoras conceived this by drawing squares of side AC, AB and BC on the three sides of a right angled triangle (►Fig. 1.3).

The trigonometric identity corresponding to Pythagoras' Theorem can be derived:

$$
\begin{aligned}
o^2 + a^2 &= h^2 \\
(\frac{o}{h})^2 + (\frac{a}{h})^2 &= 1 \\
\sin^2 \theta + \cos^2 \theta &= 1
\end{aligned}
$$

Similarly, $\sin^2 \phi + \cos^2 \phi = 1$.

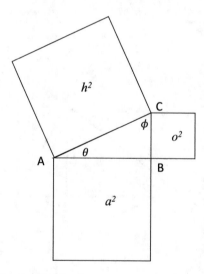

□ **Fig. 1.3** Diagram of Pythagoras' Theorem

1

1.3 The Sine and Cosine Rules

1.3.1 Sine Rule

The $\perp$ from C to BA in the general triangle of Fig. 1.4 produces two right angled triangles. Then,

$$\sin B = \frac{h}{a} \Rightarrow h = a\sin B = b\sin A$$

$$\frac{\sin A}{a} = \frac{\sin B}{b}$$

In an identical fashion, we could drop the $\perp$ from B onto AC and prove that

$$\frac{\sin C}{c} = \frac{\sin A}{a}$$

which gives the complete form of the Sine Rule:

$$\frac{\sin A}{a} = \frac{\sin B}{b} = \frac{\sin C}{c}$$

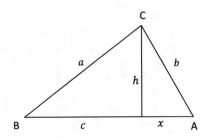

◘ **Fig. 1.4** General triangle

1.3.2 **Cosine Rule**

This is a useful rule in trigonometry when the triangle has no right angle and pairs of angle and opposite sides are not known.
Then,

$$a^2 = b^2 + c^2 - 2bc \cos A$$

Proof from ▶Fig. 1.4:

$$
\begin{aligned}
a^2 &= h^2 + (c - x)^2 \\
b^2 &= h^2 + x^2 \\
\therefore \quad a^2 &= b^2 + c^2 - 2cx \\
x &= b \cos A \\
\Rightarrow \quad a^2 &= b^2 + c^2 - 2bc \cos A \quad \text{Q.E.D.}
\end{aligned}
$$

1.4 **Compound Angles**

For any two angles A, B, from ▶Fig. 1.5

$$
\begin{aligned}
\sin(A + B) &= \frac{MP}{OP} = \frac{MQ + QP}{OP} = \frac{MQ}{OP} + \frac{QP}{OP} \\
&= \frac{RN}{ON} \cdot \frac{ON}{OP} + \frac{QP}{PN} \cdot \frac{PN}{OP} \\
\sin(A + B) &= \sin A \cos B + \cos A \sin B \\
\cos(A + B) &= \frac{OM}{OP} = \frac{OR - MR}{OP} = \frac{OR - QN}{OP} \\
&= \frac{OR}{ON} \cdot \frac{ON}{OP} - \frac{QN}{PN} \cdot \frac{PN}{OP} \\
\cos(A + B) &= \cos A \cos B - \sin A \sin B
\end{aligned}
$$

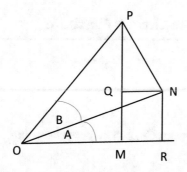

◘ **Fig. 1.5** Triangles constructed for compound angle $A + B$

1

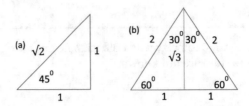

■ **Fig. 1.6**   Useful triangles

Calculators are not necessary when working with certain angles, and dispensing with them promotes better understanding.

For example, from ▶Fig. 1.6a

$$\sin 45° \;=\; \cos 45° = \frac{1}{\sqrt{2}}$$

$$\tan 45° \;=\; 1$$

From ▶Fig. 1.6b

$$\sin 60° \;=\; \frac{\sqrt{3}}{2} = \cos 30°$$

$$\tan 60° \;=\; \sqrt{3}$$

$$\tan 30° \;=\; \frac{1}{\sqrt{3}}$$

Example: Calculate $\sin 75°$

$$
\begin{aligned}
\sin 75° \;&=\; \sin(45° + 30°) \\
&=\; \sin 45° \cos 30° + \cos 45° \sin 30° \\
&=\; \frac{1}{\sqrt{2}} \cdot \frac{\sqrt{3}}{2} + \frac{1}{\sqrt{2}} \cdot \frac{1}{2} = \frac{\sqrt{3}+1}{2\sqrt{2}} = \frac{\sqrt{6}+\sqrt{2}}{4}
\end{aligned}
$$

1.5   Derivatives of the Circular Functions

$$y \;=\; \sin x$$

$$\frac{\mathrm{d}y}{\mathrm{d}x} \;=\; \cos x$$

Students may wish to consult Chap. 5 before proceeding.

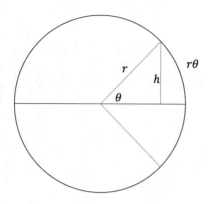

☐ **Fig. 1.7** Definition of an Angle

$$\frac{dy}{dx} = \lim_{\delta x \to 0} \frac{y(x + \delta x) - y(x)}{\delta x}$$

$$= \lim_{\delta x \to 0} \left[\frac{\sin(x + \delta x) - \sin x}{\delta x} \right]$$

$$= \lim_{\delta x \to 0} \left[\frac{\sin x \cos \delta x + \cos x \sin \delta x - \sin x}{\delta x} \right]$$

$$\simeq \lim_{\delta x \to 0} \left[\frac{\sin x + \cos x \delta x - \sin x}{\delta x} \right]$$

$$= \cos x$$

We have used the approximation $\sin \delta x \simeq \delta x$ since δx is small. This is clear from ▶Fig. 1.7 for any angle θ. Arc length $a = r\theta \simeq h$ for small θ ∴ $r\theta \simeq r \sin \theta \Rightarrow \theta \simeq \sin \theta$.

Exercise:

1. Show that if $y = \cos x$, $\dfrac{dy}{dx} = -\sin x$

2. Show that if $y = \tan x$, $\dfrac{dy}{dx} = \sec^2 x$

Note: In functional analysis, the units of x are determined by the context and may be metres, kilograms, ms^{-1}, etc. But when $f(x)$ is a trigonometric function like $\sin x$, the units of x must be units of angle: degrees or radians.

Real and Complex Numbers

Contents

© The Author(s), under exclusive license to Springer Nature
Switzerland AG 2022
P. Prewett, *Foundation Mathematics for Science and Engineering Students*,
https://doi.org/10.1007/978-3-030-91963-4_2

2

Real numbers can lie anywhere on the Real Line $\mathbb{R}$ between $-\infty$ and ∞. We write $x \in \mathbb{R}$, $-\infty < x < \infty$ (▶Fig. 2.1).

Integers are whole numbers and can be positive or negative, e.g. 1, 2, 3. Rational real numbers can be expressed as the ratio of 2 integers. e.g $1.5 \equiv \frac{3}{2}$ is rational. $\pi = 3.1415926\ldots$ is irrational, but not all endless decimals are irrational. e.g. $0.333 \cdots \equiv \frac{1}{3}$ is rational.

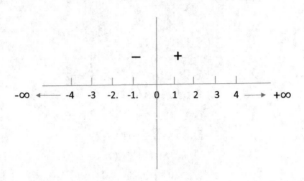

▢ Fig. 2.1 Schematic of the Real Line $\mathbb{R}$

2.1 Surds

These number forms frequently occur in applications. Their general form is $x \pm \sqrt{y}$ where $x, y \in \mathbb{R}$. When the root occurs in the denominator, the conjugate surd can be used to simplify the expression.

e.g.

$$\frac{2}{\sqrt{3} - 1}$$

Multiplying numerator and denominator by the conjugate $\sqrt{3} + 1$ gives

$$\frac{2}{\sqrt{3} - 1} = \frac{2(\sqrt{3} + 1)}{2} = \sqrt{3} + 1 = 2.732$$

Show that $\dfrac{3}{\sqrt{2} + 1} = 1.242$ using the conjugate method.

The conjugate multiplication method is also commonly used in the theory of complex numbers (see ▶Sect. 2.3).

2.2 Solution of Quadratic Equations

The general form of a quadratic equation is

$$ax^2 + bx + c = 0$$

Consider, for the moment, only equations having real roots; the case of quadratic equations having purely imaginary or complex roots is dealt with in ▶Sect. 2.3. Some equations with real roots can be factorized and others can't. Each case will be dealt with by example.

2.2.1 Equations Which Can Be Factorized

$$
\begin{aligned}
x^2 + 3x - 10 &= 0 \\
(x - 2)(x + 5) &= 0 \\
\therefore x - 2 = 0 \quad &\text{or} \quad x + 5 = 0
\end{aligned}
$$

so that the equation has 2 real roots $x = 2, -5$
Another example:

$$
\begin{aligned}
x^2 + 10x + 24 &= 0 \\
(x + 4)(x + 6) &= 0
\end{aligned}
$$

$\therefore x = -4, -6$ are the two roots.

Quadratic equations may involve perfect squares, e.g.

$$
\begin{aligned}
x^2 - 6x + 9 &= 0 \\
(x - 3)^2 &= 0 \\
x &= 3
\end{aligned}
$$

This type of equation has only one root, sometimes called repeated roots.

2.2.2 Equations Which Cannot Be Factorized

$$x^2 + 3x - 5 = 0$$

From the general form

$$x^2 + \frac{bx}{a} + \frac{c}{a} = 0$$

2

$$(x + \frac{b}{2a})^2 - \frac{b^2}{4a^2} + \frac{c}{a} = 0$$

$$x + \frac{b}{2a} = \pm\frac{\sqrt{b^2 - 4ac}}{2a}$$

The 2 roots are therefore

$$x_1, x_2 = \frac{-b \pm \sqrt{b^2 - 4ac}}{2a}$$

Example:

$$x^2 + 3x - 5 = 0$$

$$x_1, x_2 = \frac{-3 \pm \sqrt{9 + 20}}{2}$$

$$x_1, x_2 = \frac{-3 \pm \sqrt{29}}{2}$$

Another example:

$$2x^2 + 7x - 3 = 0$$

$$x_1, x_2 = \frac{-7 \pm \sqrt{49 + 24}}{2}$$

$$x_1, x_2 = \frac{-7 \pm \sqrt{73}}{2}$$

If the term $b^2 - 4ac$ has an integer square root, the equation can be factorized and that method should be used.
Example:

$$x^2 + 10x + 21 = 0$$

$$\sqrt{b^2 - 4ac} = \sqrt{100 - 84} = \sqrt{16} = 4$$
$$(x + 7)(x + 3) = 0 \Rightarrow x_1, x_2 = -3, -7.$$

2.3 Complex Numbers

$x^2 = 1$ has solutions $x = \pm 1$. But if $x^2 = -1$, there are no real solutions for which $x \in \mathbb{R}$.
In this case there is just one imaginary number root, written $x = i$ where $i = \sqrt{-1}$.

$x^2 = -9$ has the single root $x = \sqrt{-9} = \sqrt{9} \cdot \sqrt{-1} = 3i$.

This rather strange concept has many important uses in science and engineering, as will be seen later.

Complex numbers have both real and imaginary parts:

$$z = 3 + 2i$$

is a complex number with real part 3 and imaginary part 2i. We can write

$$\begin{aligned} \text{Re}(z) &= 3 \\ \text{Im}(z) &= 2i \end{aligned}$$

Complex numbers may occur when we solve quadratic equations (see ▶Sect. 2.2). As seen earlier, the quadratic equation

$$ax^2 + bx + c = 0$$

has solutions

$$x_1, x_2 = \frac{-b \pm \sqrt{b^2 - 4ac}}{2a}$$

If $b^2 > 4ac$, the square root term is real and the 2 roots of the equation are also real

$$x_1, x_2 \in \mathbb{R}$$

But what if $b^2 < 4ac$ as for the quadratic equation

$$z^2 - 2z + 6 = 0?$$

Then,

$$z_1, z_2 = 1 \pm \sqrt{-5}$$

The roots $z_1, z_2 = 1$ are complex numbers composed of a real part and an imaginary part. Thus

$$z_1, z_2 \in \mathbb{C}$$

where $\mathbb{C}$ is the Complex Plane.

The Real and imaginary parts of the complex roots are

$$\begin{aligned} \text{Re}(z_1) &= 1; & \text{Im}(z_1) &= i\sqrt{5} \\ \text{Re}(z_2) &= 1; & \text{Im}(z_2) &= -i\sqrt{5} \end{aligned}$$

They can be represented as points on $\mathbb{C}$, as in ▶Fig. 2.2.

Any complex number can be represented by a point in the complex plane. In this example, z_1 and z_2 are complex conjugates and are frequently written as $z, \bar{z}$ or z, z^*.

2

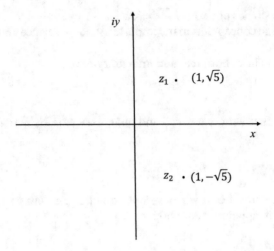

□ Fig. 2.2   Pair of complex roots $z_1, z_2 = 1 \pm i\sqrt{5}$ represented by two conjugate points in the complex plane

In general $z = x + iy$ is represented by the point (x, y) in $\mathbb{C}$.

$$
\begin{aligned}
z &= x + iy \\
\text{and } \bar{z} &= x - iy
\end{aligned}
$$

$$
z\bar{z} = (x + iy)(x - iy) = x^2 + y^2 = |z|^2
$$

where $|z|$ is the modulus of z. z is completely defined by polar coordinates $|z|$ and θ. The conjugate pair differ only in their angle in the Argand diagram ($\pm\theta$ in Fig. 2.3). i.e. $\theta_{\bar{z}} = (2\pi - \theta_z)$. Their moduli are indentical ($|\bar{z}| = |z|$).

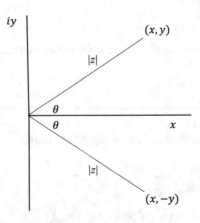

□ Fig. 2.3   Argand representation of complex numbers

2.4 Euler's Theorem

This is a key theorem for science and engineering, being used in the theory of waves, electrical circuit theory and quantum mechanics.

$$e^{i\theta} = \cos\theta + i\sin\theta$$

[1] The McLaurin Series (see ►Sect. 7.5) gives

$$
\begin{aligned}
\cos\theta &= 1 - \frac{\theta^2}{2!} + \frac{\theta^4}{4!} - \frac{\theta^6}{6!} + \cdots \\
\sin\theta &= \frac{\theta}{1!} - \frac{\theta^3}{3!} + \frac{\theta^5}{5!} - \cdots \\
e^{\theta} &= 1 + \frac{\theta}{1!} + \frac{\theta^2}{2!} + \frac{\theta^2}{3!} + \cdots \\
e^{i\theta} &= 1 + \frac{i\theta}{1!} - \frac{\theta^2}{2!} - \frac{i\theta^3}{3!} + \frac{\theta^4}{4!} + \frac{i\theta^5}{5!} - \cdots
\end{aligned}
$$

Therefore

$$e^{i\theta} = \cos\theta + i\sin\theta \quad \text{Q.E.D.}$$

2.4.1 De Moivre's Theorem

$$e^{in\phi} = \cos n\phi + i\sin n\phi, \quad \text{integer } n$$

This is essentially the same as Euler's Formula, which is obtained from it by putting $n\phi = \theta$.

With θ as drawn in ►Fig. 2.3, the Argand Diagram can be seen to represent $z \in \mathbb{C}$ which can be expressed as

$$z = |z|e^{i\theta}$$

This is often the most useful way of writing a complex number z.

This also gives an easier way of proving trigonometric identities than the complex geometrical methods in ►Sect. 1.4.

Thus, for the compound angle formulae of ►Sect. 1.5,

$$
\begin{aligned}
z = e^{i(A+B)} &= e^{iA} \cdot e^{iB} \\
&= (\cos A + i\sin A)(\cos B + i\sin B) \\
e^{i(A+B)} &= \cos(A+B) + i\sin(A+B) \\
&= \cos A \cos B - \sin A \sin B \\
&\quad + i(\sin A \cos B + \cos A \sin B)
\end{aligned}
$$

---

[1] See page 75 for Euler's Number e, and ►Sect. 7.5.1 for series expansions of e^{θ}, $\sin\theta$ and $\cos\theta$.

2

Equating real and imaginary parts,

$$\text{Re}(z): \quad \cos(A+B) \; = \; \cos A \cos B - \sin A \sin B$$
$$\text{Im}(z): \quad \sin(A+B) \; = \; \sin A \cos B + \cos A \sin B$$

Example: Use de Moivre's Theorem to show that

$$\sin^5 \theta = \frac{1}{2^4}(\sin 5\theta - 5\sin 3\theta + 10\sin\theta)$$

$$\sin^5\theta \quad = \quad \left[\frac{e^{i\theta}-e^{-i\theta}}{2i}\right]^5 = \frac{e^{i5\theta}}{2^5 i}(1-e^{-i2\theta})^5$$

Using   the   Binomial Theorem - see ▶Sect. 7.4

$$(1+X)^5 \quad = \quad 1 + 5X + \tfrac{5\cdot4}{1\cdot2}X^2 + \tfrac{5\cdot4\cdot3}{1\cdot2\cdot3}X^3 + \tfrac{5\cdot4\cdot3\cdot2}{1\cdot2\cdot3\cdot4}X^4 + X^5$$

where $X \quad = \quad -e^{-i2\theta}$

$$\sin^5\theta \quad = \quad \frac{e^{i5\theta}}{2^5 i}\left[1 - 5e^{-i2\theta} + 10e^{-i4\theta} - 10e^{-i6\theta} + 5e^{-i8\theta} - 10e^{-i4\theta}\right]$$

$$\sin^5\theta \quad = \quad \frac{1}{2^5 i}\left[e^{i5\theta} - 5e^{i3\theta} + 10e^{i\theta} - 10e^{-i\theta} + 5e^{-i3\theta} - e^{-i5\theta}\right]$$

$$= \quad \frac{1}{2^5 i}\left[(e^{i5\theta} - e^{-i5\theta}) - 5(e^{i3\theta} - e^{-i3\theta}) + 10(e^{i\theta} - e^{-i\theta})\right]$$

$$\sin^5\theta \quad = \quad \tfrac{1}{2^4}[\sin 5\theta - 5\sin 3\theta + 10\sin\theta] \quad \text{Q.E.D.}$$

2.5 Phase Ambiguity

$|z| = \sqrt{x^2 y^2}$ is uniquely defined but the phase $\arg(z)$ is not.
We usually take $\arg(z)$ to be θ where $\tan\theta = \dfrac{y}{x}$ but

$\theta_n = \tan^{-1}\dfrac{y}{x} + n2\pi$ for any integer n will also give the same complex z. This has consequences for $f(z)$ such as $\ln z$. Thus, $\ln z = \ln|z| + i(\theta + 2\pi n)$. Different values of n represent different "branches" of the function $\ln z$. $n = 0$ is the Principal Branch of $\ln z$.
For a product of complex numbers,

$$\ln(z_1 z_2) \quad = \quad \ln(|z_1||z_2|e^{i(\theta_1+\theta_2)})$$
$$= \quad \ln|z_1| + \ln|z_2| + i(\theta_1 + \theta_2 + n2\pi + m2\pi)$$
$$= \quad \ln|z_1| + \ln|z_2| + i(\theta_1 + \theta_2 + k2\pi)$$

where integer $k = n + m$ depends on which branch of z_1 and z_2 is chosen ($n = 0, m = 0$ are the Principal Branches of $\ln z_1$ and $\ln z_2$ respectively).
The logarithm of the product $z_1 z_2$ has a multiplicity of phases, varying by $2\pi k$ where each integer $0 \le k \le \infty$ represents a different branch of the solution.

Vector Algebra

Contents

© The Author(s), under exclusive license to Springer Nature
Switzerland AG 2022
P. Prewett, *Foundation Mathematics for Science and Engineering Students*,
https://doi.org/10.1007/978-3-030-91963-4_3

3

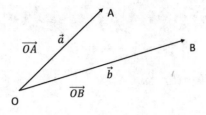

□ **Fig. 3.1**   Vector quantities

A vector has magnitude (length of OA) and direction. $\overrightarrow{OA}$ and $\overrightarrow{OB}$ have the same magnitude but different direction, hence (►Fig. 3.1)

$$\mathbf{a} \neq \mathbf{b}$$

3.1 Addition of Vectors

Referring to ►Fig. 3.2a, think in terms of displacement. Walk from O to A as represented by the magnitude of $\overrightarrow{OA}$ and its direction, i.e. by vector $\mathbf{a}$. Then walk from A to C (vector $\mathbf{b}$) to arrive at C. This is just the same as taking the path OC represented by $\mathbf{c}$, giving the Triangle of Vectors Law of Addition

$$\mathbf{c} = \mathbf{a} + \mathbf{b}$$

The vectors must be drawn "nose to tail" as shown.
Completing the Parallelogram of Vectors reveals the second triangle and shows that

$$\mathbf{c} = \mathbf{a} + \mathbf{b} = \mathbf{b} + \mathbf{a}$$

It is evident, therefore, that vector addition is commutative and there are an infinity of triangles of vectors for each vector $\mathbf{c}$. $\mathbf{c}$ is the *resultant* vector of vectors $\mathbf{a}$ and $\mathbf{b}$, $\mathbf{d}$ and $\mathbf{e}$, $\mathbf{f}$ and $\mathbf{g}$, $\mathbf{p}$ and $\mathbf{q}$, as drawn in ►Fig. 3.2b.
Any number of vectors may be added in nose-to-tail configuration to form a Polygon of Vectors (►Fig. 3.2c) for which the resultant is $\mathbf{R} = \sum_{r=1}^{n} \mathbf{a}_r$.

Every vector has magnitude (modulus $|\mathbf{c}|$) and direction.
Choosing two mutually perpendicular directions to describe a vector using the triangle of vectors, is shown in ►Fig. 3.3:

$$\mathbf{c} = \mathbf{a} + \mathbf{b} = |a|\,\mathbf{i} + |b|\,\mathbf{j}$$

where $\mathbf{i}, \mathbf{j}$ are *unit vectors* in the mutually orthogonal directions $\overrightarrow{OA}$ and $\overrightarrow{AB}$ (same as $\overrightarrow{CB}$ and $\overrightarrow{OC}$ from the diagram).

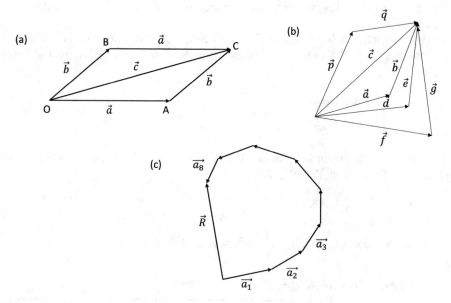

□ Fig. 3.2 **a** Sum of two vectors, **b** an infinity of paths, **c** Polygon of vectors

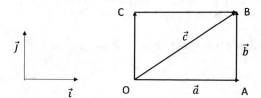

□ Fig. 3.3 Vector resolved into two ⊥ vectors

It is usual to drop the modulus notation and write

$$\mathbf{c} = a\mathbf{i} + b\mathbf{j}$$

Because orthogonal vectors have been chosen, giving a right angled triangle of vectors,

$$c = \sqrt{a^2 + b^2}$$

and vector **c** may be written

$$\mathbf{c} = (\sqrt{a^2 + b^2})\hat{\mathbf{c}} = a\mathbf{i} + b\mathbf{j}$$

where $\hat{\mathbf{c}}$ is the unit vector in direction $\overrightarrow{OB}$.

3

This process of splitting vectors into orthogonal component vectors is called Resolution of Vectors.

The above analysis is for vectors defined on a plane, but it can obviously be extended to 3-dimensional vectors by introducing a third unit vector **k**, orthogonal to **i** and **j**, giving

$$\mathbf{a} = a_x\mathbf{i} + a_y\mathbf{j} + a_z\mathbf{k}$$
$$a = \sqrt{a_x^2 + a_y^2 + a_z^2}$$

We now have an algebraic method of adding vectors, alongside the geometrical method of Triangle of Vectors, thus:

$$\mathbf{a} = a_x\mathbf{i} + a_y\mathbf{j} + a_x\mathbf{k}$$
$$\mathbf{b} = b_x\mathbf{i} + b_y\mathbf{j} + b_z\mathbf{k}$$
$$\mathbf{c} = \mathbf{a} + \mathbf{b} = (a_x + b_x)\mathbf{i} + (a_y + b_y)\mathbf{j} + (a_z + b_z)\mathbf{k}$$

3.2 Subtraction of Vectors

Vectors **a** and $-\mathbf{a}$ in ▶Fig. 3.4 have the same magnitude and direction, but opposite sign. Adding them is the same as subtraction and gives the zero vector $\mathbf{a} + (-\mathbf{a}) = \mathbf{a} - \mathbf{a} = \underline{0}$. For any two vectors **a** and **b**,

$$\mathbf{a} - \mathbf{b} = (a_x - b_x)\mathbf{i} + (a_y - b_y)\mathbf{j} + (a_z - b_z)\mathbf{k}$$

□ **Fig. 3.4** Sign of vector (subtraction)

3.3 Multiplication of Vectors

Their are two different products of vectors: scalar product and vector product. Scalar product $\mathbf{a} \cdot \mathbf{b} = ab \cos \theta$ is not a vector. It is a scalar formed from 2 vectors, containing information about their relative directions ($\cos \theta$) and their magnitudes $a \cdot b$, see ►Fig. 3.5.

The special case when $\mathbf{a}$ and $\mathbf{b}$ are unit vectors $\mathbf{i}$ and $\mathbf{j}$ gives

$$\mathbf{i} \cdot \mathbf{j} = |\mathbf{i}||\mathbf{j}| \cos \frac{\pi}{2} = 0$$
$$\mathbf{i} \cdot \mathbf{i} = |\mathbf{i}||\mathbf{i}| \cos 0 = 1$$

This "orthogonality" is true for the set of 3 unit vectors $\mathbf{i}, \mathbf{j}, \mathbf{k}$ in 3 dimensions. The scalar product $\mathbf{a} \cdot \mathbf{b} = ab \cos \theta$ can also be written as $\mathbf{a} \cdot \mathbf{b} = a_x b_x + a_y b_y + a_z b_z$. Note that the scalar product operation is commutative

$$\mathbf{a} \cdot \mathbf{b} = ab \cos \theta = ba \cos \theta = \mathbf{b} \cdot \mathbf{a}$$

This may also be written, for a 2-D vector, as

$$\mathbf{a} \cdot \mathbf{b} = a_x b_x + a_y b_y$$

The proof of this is as follows, referring to ►Fig. 3.5:

$$b_x = b \cos(\theta + \phi)$$
$$b_y = b \sin(\theta + \phi)$$

$$b_x = b(\cos \theta \cos \phi - \sin \theta \sin \phi)$$
$$b_y = b(\sin \theta \cos \phi + \cos \theta \sin \phi)$$
$$a_x = a \cos \phi$$
$$a_y = a \sin \phi$$

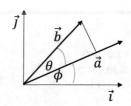

◘ **Fig. 3.5** Relative directions of two vectors

$$b_x = b\frac{a_x}{a}\cos\theta - b\frac{a_y}{a}\sin\theta \qquad (3.1)$$

$$b_y = b\frac{a_y}{a}\cos\theta + b\frac{a_x}{a}\sin\theta \qquad (3.2)$$

Multiplying ▶Eq. (3.1) by a_x and ▶Eq. (3.2) by a_y and eliminating $\sin\theta$ shows

$$a_xb_x + a_yb_y = ab\cos\theta \qquad (3.3)$$

This is readily extended to 3D vectors:

$$\mathbf{a}\cdot\mathbf{b} = ab\cos\theta = a_xb_x + a_yb_y + a_zb_z$$

3.3.1 Mutually ⊥ Vectors

Such vectors will satisfy $\mathbf{a}\cdot\mathbf{b} = 0$
($\mathbf{a}\cdot\mathbf{b} = ab\cos\theta = 0 \Rightarrow \cos\theta = 0, \theta = \frac{\pi}{2}$).
For example, vectors $2\mathbf{i} + 6\mathbf{j}$ and $3\mathbf{i} - \mathbf{j}$ are mutually perpendicular since

$$\mathbf{a}\cdot\mathbf{b} = a_xb_x + a_yb_y = 2\times 3 + 6\times(-1) = 0$$

Also vectors $\mathbf{a} = \mathbf{i} - 2\mathbf{j}$ and $\mathbf{b} = 4\mathbf{i} + 2\mathbf{j}$ are mutually ⊥ since $\mathbf{a}\cdot\mathbf{b} = 1\times 4 - 2\times 2 = 0$
Find the vector perpendicular to vector $\mathbf{a} = 3\mathbf{i} + 6\mathbf{j}$. This will be $\mathbf{b}$ such that

$$3b_x + 6b_y = 0$$

There is an infinite number of solutions for $\mathbf{b} \perp \mathbf{a}$ which satisfy this equation. For example

1. $\mathbf{i} - \frac{1}{2}\mathbf{j}$
2. $2\mathbf{i} - \mathbf{j}$
3. $3\mathbf{i} - \frac{3}{2}\mathbf{j}$

Check:

1. $(3\mathbf{i} + 6\mathbf{j})\cdot(\mathbf{i} - \frac{1}{2}\mathbf{j}) = 3 - \frac{1}{2}\times 6 = 0$
2. $(2\mathbf{i} - \mathbf{j})\cdot(3\mathbf{i} + 6\mathbf{j}) = 6 - 6 = 0$
3. $(3\mathbf{i} - \frac{3}{2}\mathbf{j})\cdot(3\mathbf{i} + 6\mathbf{j}) = 9 - \frac{3}{2}\times 6 = 0$

All of these vectors have the same direction and are scalar multiples of each other. All vectors which are scalar multiples of ⊥ vectors will also be ⊥ to each other. Note that multiplication by a scalar is not the same as scalar product.

$$\lambda \mathbf{a} = \lambda(a_x\mathbf{i} + a_y\mathbf{j} + a_z\mathbf{k})$$
$$= \lambda a_x\mathbf{i} + \lambda a_y\mathbf{j} + \lambda a_z\mathbf{k}$$

Vectors $\lambda\mathbf{a}$ and $\mathbf{a}$ have the same direction but their magnitudes differ by the factor λ.

$$|\mathbf{a}| = \sqrt{a_x^2 + a_y^2 + a_z^2}$$
$$|\lambda\mathbf{a}| = \sqrt{(\lambda a_x)^2 + (\lambda a_y)^2 + (\lambda a_z)^2}$$
$$= \lambda|\mathbf{a}|$$

3.4 Direction Cosines

The direction in which any vector $\mathbf{a}$ points can be specified by the angles α, β, γ it makes with the Cartesian axes (i.e. with the orthogonal set $\mathbf{i}, \mathbf{j}, \mathbf{k}$ of unit vectors). The Direction Cosines of $\mathbf{a}$ are the cosines of these three angles (▶Fig. 3.6).

$$\cos\alpha = \frac{a_x}{a}$$
$$\cos\beta = \frac{a_y}{a}$$
$$\cos\gamma = \frac{a_z}{a}$$

Unit vector in the $\mathbf{a}$-direction is $\hat{\mathbf{a}} = \dfrac{\mathbf{a}}{a}$

Direction cosines of $\lambda\mathbf{a}$ are

$$\cos\alpha' = \frac{\lambda a_x}{\lambda a} = \frac{a_x}{a} = \cos\alpha$$

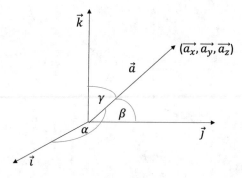

☐ **Fig. 3.6** Direction cosines

3

and the same for β and γ, so that $\lambda\mathbf{a}$ and $\mathbf{a}$ are shown to be in the same direction. What are the direction cosines of the vector $\mathbf{a} = (1, 1, 0)$?

$$|\mathbf{a}| = \sqrt{1^2 + 1^2 + 0^2} = \sqrt{2}$$

i.e. $\cos\alpha = \dfrac{a_x}{|\mathbf{a}|} = \dfrac{1}{\sqrt{2}};\quad \cos\beta = \dfrac{1}{\sqrt{2}};\quad \cos\gamma = 0$

Therefore $\alpha = 45°$, $\beta = 45°$, $\gamma = \frac{\pi}{2}$.

3.5 Position Vector r

$\mathbf{r} = x\mathbf{i} + y\mathbf{j} + z\mathbf{k}$ is the position vector of $P(x, y, z)$ ($\blacktriangleright$Fig. 3.7). To obtain the velocity vector requires differentiation of the position vector:

$$\mathbf{v} = \frac{d\mathbf{r}}{dt} = \dot{x}\mathbf{i} + \dot{y}\mathbf{j} + \dot{z}\mathbf{k}$$

provided the axes are fixed in space and time. (Sometimes this is not the case). As for $\mathbf{r}$, for any vector $\mathbf{a}$

$$\begin{aligned}\frac{d\mathbf{a}}{dt} &= \frac{d}{dt}(a_x\mathbf{i} + a_y\mathbf{j} + a_z\mathbf{k})\\ &= \dot{a}_x\mathbf{i} + \dot{a}_y\mathbf{j} + \dot{a}_z\mathbf{k}\end{aligned}$$

If the position vectors of two points, A and B are known relative to an arbitrary origin, the vector position of A relative to B is simply calculated from the triangle of vectors as shown in $\blacktriangleright$Fig. 3.8.

$$\vec{x}_{AB} = \vec{x}_B - \vec{x}_A = (6, 2, 3)$$

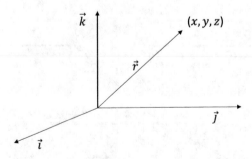

☐ **Fig. 3.7** Position vector

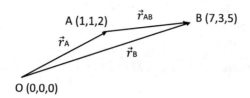

◻ Fig. 3.8   Relative position vector

3.6 **Vector/Cross Product**

$\mathbf{a} \wedge \mathbf{b} = ab \sin \theta \hat{\mathbf{e}}$ where $\hat{\mathbf{e}}$ is the unit vector perpendicular to the plane containing **a** and **b**. $\hat{\mathbf{e}}$ is generated in the right hand corkscrew sense, twisting from **a** to **b**.

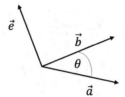

◻ Fig. 3.9   Vector cross product - right hand corkscrew rule

$-\hat{\mathbf{e}}$ points in the opposite direction to $\hat{\mathbf{e}}$ i.e. twisting from **b** to **a**.
Clearly, $\mathbf{a} \wedge \mathbf{b} = -\mathbf{b} \wedge \mathbf{a}$, which shows that the vector product operator is not commutative since $\mathbf{a} \wedge \mathbf{b} \neq \mathbf{b} \wedge \mathbf{a}$. (To distinguish the vector product from scalar multiplication, we prefer to use $\wedge$ rather than $\times$) (►Fig. 3.9).

For unit vectors **i**, **j**, **k**
$$\begin{cases} \mathbf{i} \wedge \mathbf{i} = 0 & \mathbf{i} \wedge \mathbf{j} = \mathbf{k} & \mathbf{j} \wedge \mathbf{i} = -\mathbf{k} \\ \mathbf{j} \wedge \mathbf{j} = 0 & \mathbf{j} \wedge \mathbf{k} = \mathbf{i} & \mathbf{k} \wedge \mathbf{j} - -\mathbf{i} \\ \mathbf{k} \wedge \mathbf{k} = 0 & \mathbf{k} \wedge \mathbf{i} = \mathbf{j} & \mathbf{i} \wedge \mathbf{k} = -\mathbf{j} \end{cases}$$

3.6.1 **Cartesian Form of Vector Product**

$$\begin{aligned} \mathbf{a} \wedge \mathbf{b} &= (a_x\mathbf{i} + a_y\mathbf{j} + a_z\mathbf{k}) \wedge (b_x\mathbf{i} + b_y\mathbf{j} + b_z\mathbf{k}) \\ &= a_x\mathbf{i} \wedge (b_x\mathbf{i} + b_y\mathbf{j} + b_z\mathbf{k}) + a_y\mathbf{j} \wedge (b_x\mathbf{i} + b_y\mathbf{j} + b_z\mathbf{k}) \\ &\quad + a_z\mathbf{k} \wedge (b_x\mathbf{i} + b_y\mathbf{j} + b_z\mathbf{k}) \\ &= a_xb_y\mathbf{k} + a_xb_z(-\mathbf{j}) + a_yb_x(-\mathbf{k}) + a_yb_z\mathbf{i} + a_zb_x\mathbf{j} \\ &\quad + a_zb_y(-\mathbf{i}) \\ \mathbf{a} \wedge \mathbf{b} &= (a_yb_z - a_zb_y)\mathbf{i} + (a_zb_x - a_xb_z)\mathbf{j} + (a_xb_y - a_yb_x)\mathbf{k} \end{aligned}$$

Therefore

$$\mathbf{a} \wedge \mathbf{b} = \begin{vmatrix} \mathbf{i} & \mathbf{j} & \mathbf{k} \\ a_x & a_y & a_z \\ b_x & b_y & b_z \end{vmatrix}$$

[The determinant form is discussed further in ►Chap. 4].
Example:

$$
\begin{aligned}
\mathbf{a} &= 2\mathbf{i} + 5\mathbf{j} - 3\mathbf{k} \\
\mathbf{b} &= \mathbf{i} - \mathbf{j} \\
\mathbf{a} \wedge \mathbf{b} &= \begin{vmatrix} \mathbf{i} & \mathbf{j} & \mathbf{k} \\ 2 & 5 & -3 \\ 1 & -1 & 0 \end{vmatrix} \\
\mathbf{a} \wedge \mathbf{b} &= \mathbf{i}[5 \cdot 0 - (-1) \cdot (-3)] - \mathbf{j}[2 \cdot 0 - (1) \cdot (-3)] \\
&\quad + \mathbf{k}[2 \cdot (-1) - 1 \cdot 5] \\
\mathbf{a} \wedge \mathbf{b} &= -3\mathbf{i} - 3\mathbf{j} - 7\mathbf{k}
\end{aligned}
$$

Orthogonality check: $\mathbf{a} = (2, 5, -3)$; $\mathbf{b} = (1, -1, 0)$
$$\mathbf{a} \wedge \mathbf{b} = (-3, -3, -7)$$

$$
\begin{aligned}
\mathbf{a} \cdot \mathbf{a} \wedge \mathbf{b} &= 2 \cdot (-3) + 5 \cdot (-3) + (-3) \cdot (-7) \\
&= -6 - 15 + 21 = 0 \\
\mathbf{b} \cdot \mathbf{a} \wedge \mathbf{b} &= 1 \cdot (-3) + (-1) \cdot (-3) + 0 \cdot (-7) \\
&= -3 + 3 = 0
\end{aligned}
$$

$\mathbf{a} \wedge \mathbf{b}$ is $\perp \mathbf{a}$ and also $\perp \mathbf{b}$ so $\mathbf{a} \wedge \mathbf{b}$ is $\perp$ to the plane containing $\mathbf{a}$ and $\mathbf{b}$.

3.7 Applications of Vectors

3.7.1 Work Done by a Force on a Body

In physics, the work done by a force pushing a body is the force × the distance moved by body in the direction of the force (►Fig. 3.10) (see ►Fig. 3.11).

$$W = (F \cos \theta) \Delta r = \mathbf{F} . \Delta \mathbf{r}$$

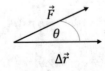

■ **Fig. 3.10** Force at angle θ to direction of motion

Example: $\mathbf{F} = 3\mathbf{i} + 7\mathbf{k}$ (units Newtons) moves a body from position A to position B in ▶Fig. 3.8.

The distance moved is $\Delta\mathbf{r} = \mathbf{r}_B - \mathbf{r}_A$ (units metres).

$$
\begin{aligned}
\Delta\mathbf{r} &= (7-1)\mathbf{i} + (3-1)\mathbf{j} + (5-2)\mathbf{k} \\
\Delta\mathbf{r} &= 6\mathbf{i} + 2\mathbf{j} + 3\mathbf{k} \\
W = \mathbf{F}\cdot\Delta\mathbf{r} &= (3\mathbf{i} + 7\mathbf{k})\cdot(6\mathbf{i} + 2\mathbf{j} + 3\mathbf{k}) \\
&= 3\times 6 + 0\times 2 + 7\times 3 \\
W &= 39\,\mathrm{J}
\end{aligned}
$$

W has units of energy, namely Joules ($\mathrm{J} \equiv \mathrm{Nm}$).

3.7.2 Moment of Force (Torque)

Referring to ▶Fig. 3.11, a force $\mathbf{F}$ (N) acting at a vector distance $\mathbf{r}$ (m)from an axis of rotation of a body through O, $\perp$ to the page (i.e. $\perp$ to the plane containing $\mathbf{r}$ and $\mathbf{F}$) produces a turning effect on the body. The magnitude of this moment, or torque, measured in Nm depends on the $\perp$ distance from the axis to $\mathbf{F}$:

$$M = aF = r\sin\theta F$$

which is the magnitude of the vector product $\mathbf{r} \wedge \mathbf{F}$. The moment is a vector

$$\mathbf{M} = \mathbf{r} \wedge \mathbf{F}$$

with direction $\perp$ out of the plane depicted in ▶Fig. 3.11. It is important in rigid body dynamics of rotating systems.

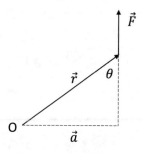

☐ **Fig. 3.11** Force acting at a distance from the axis of rotation, producing a turning moment or torque

Matrices

Contents

© The Author(s), under exclusive license to Springer Nature
Switzerland AG 2022
P. Prewett, *Foundation Mathematics for Science and Engineering Students*,
https://doi.org/10.1007/978-3-030-91963-4_4

4.1 **Basic Matrices**

A matrix is a rectangular array of real numbers, in general with n rows, m columns, but can be square ($n = m$).
Examples:

$$A = \begin{bmatrix} 4 & 1 \\ 3 & 6 \\ -1 & 0 \end{bmatrix} \qquad B = \begin{bmatrix} 1 & 0 & -1 & 3 \\ 6 & 1 & 4 & 2 \end{bmatrix}$$
$$3 \times 2 \qquad\qquad\qquad 2 \times 4$$

Sometimes written $\mathbf{A}$, $\mathbf{B}$ or $\underline{\underline{A}}$, $\underline{\underline{B}}$ and called tensors.
A matrix/tensor is a 2D array, whereas a vector is a linear array.
Examples:

$$\text{Vector } \mathbf{a} = 3\mathbf{i} + 2\mathbf{j} + 4\mathbf{k} \equiv (3, 2, 4)$$
$$\text{Tensor } \mathbf{X} = \begin{bmatrix} 1 & 2 & 3 \\ 7 & 5 & 2 \\ 3 & 2 & 4 \end{bmatrix}$$

Elements of a vector: a_i, $i = 1, 2, 3$
Elements of a Matrix/Tensor: a_{ij} where $i \in [1, n]; j \in [1, m]$
For $\mathbf{X}$ above, $x_{23} = 2$; $x_{31} = 3$

4.1.1 **Addition and Subtraction of Matrices**

$$A = \begin{bmatrix} 7 & 2 \\ -2 & 1 \\ 4 & 3 \end{bmatrix} \qquad B = \begin{bmatrix} -1 & 0 \\ 3 & 4 \\ 9 & 1 \end{bmatrix}$$

$$A + B = \begin{bmatrix} 6 & 2 \\ 1 & 5 \\ 13 & 4 \end{bmatrix} = B + A \text{ (commutative)}$$

$$A - B = \begin{bmatrix} 8 & 2 \\ -5 & -3 \\ -5 & 2 \end{bmatrix} = -B + A \text{ (commutative)}$$

Note that matrices of different sizes cannot be added:

$$\begin{bmatrix} 3 & 2 & 0 \\ 1 & 2 & 3 \\ 5 & 2 & 7 \end{bmatrix} + \begin{bmatrix} 1 & 5 & 4 \\ 2 & 3 & 1 \end{bmatrix}$$

cannot be added because there is a left over row in the 1st matrix with no matching row in the second matrix.

4.1.2 **Multiplying a Matrix by a Scalar**

$$\alpha \times A = \alpha \begin{bmatrix} 2 & 3 \\ 5 & 1 \end{bmatrix} = \begin{bmatrix} 2\alpha & 3\alpha \\ 5\alpha & \alpha \end{bmatrix}$$

Each element of the matrix is multiplied by the number α.

Thus, $\quad 3A = 3 \times \begin{bmatrix} 2 & 3 \\ 5 & 1 \end{bmatrix} = \begin{bmatrix} 6 & 9 \\ 15 & 3 \end{bmatrix}$

Where each matrix element has a common factor, it may be convenient to extract this as a scalar multiplier.
Example:

$$A = \begin{bmatrix} 2 & 8 & 6 \\ 8 & 16 & 4 \\ 12 & 10 & 14 \end{bmatrix} = 2 \begin{bmatrix} 1 & 4 & 3 \\ 4 & 8 & 2 \\ 6 & 5 & 7 \end{bmatrix}$$

Multiplication by a scalar is commutative.

$$A = 2 \times B = B \times 2$$

4.1.3 **Multiplication of Matrices**

$$\begin{bmatrix} 2 & 3 & 4 \\ 3 & 2 & 5 \end{bmatrix} \times \begin{bmatrix} 2 & 5 \\ 1 & 2 \\ 3 & 5 \end{bmatrix} = \begin{bmatrix} 19 & 36 \\ 23 & 44 \end{bmatrix}$$
$$\quad 2 \times 3 \qquad\qquad 3 \times 2 \qquad\quad 2 \times 2$$

Multiply columns of matrix 2 by rows of matrix 1.

$$[n \times m] \times [m \times n] \rightarrow [n \times n]$$

The product of 2 rectangular matrices is a square matrix.

4.1.4 **Square Matrices**

$$A = \begin{bmatrix} 1 & 3 & -1 \\ & \ddots & \\ 0 & 2 & 4 \\ & & \ddots \\ -1 & 5 & 7 \end{bmatrix}$$

4

Square matrices have a diagonal, as shown.

All square matrices have an Identity Matrix I for which the diagonal elements are 1. For example for a 3×3 matrix,

$$I_3 = \begin{bmatrix} 1 & 0 & 0 \\ 0 & 1 & 0 \\ 0 & 0 & 1 \end{bmatrix}$$

$$A \times I_3 = \begin{bmatrix} 1 & 3 & -1 \\ 0 & 2 & 4 \\ -1 & 5 & 7 \end{bmatrix} \times \begin{bmatrix} 1 & 0 & 0 \\ 0 & 1 & 0 \\ 0 & 0 & 1 \end{bmatrix} = \begin{bmatrix} 1 & 3 & -1 \\ 0 & 2 & 4 \\ -1 & 5 & 7 \end{bmatrix}$$

Clearly, A is unchanged when multiplied by the identity matrix.

For 2×2 matrices, $I_2 = \begin{bmatrix} 1 & 0 \\ 0 & 1 \end{bmatrix}$

$$\begin{bmatrix} 1 & 3 \\ 2 & 4 \end{bmatrix} \times \begin{bmatrix} 1 & 0 \\ 0 & 1 \end{bmatrix} = \begin{bmatrix} 1 & 3 \\ 2 & 4 \end{bmatrix}$$

$$\begin{bmatrix} 1 & 0 \\ 0 & 1 \end{bmatrix} \times \begin{bmatrix} 1 & 3 \\ 2 & 4 \end{bmatrix} = \begin{bmatrix} 1 & 3 \\ 2 & 4 \end{bmatrix}$$

$A \times I = I \times A$ (commutative)

N.B. The above only applies for square matrices.

4.1.5 Transpose of a Matrix A^T

A^T is obtained by switching rows and columns.

$$A = \begin{bmatrix} 1 & 3 & -1 \\ 0 & 2 & 4 \\ -1 & 5 & 7 \end{bmatrix} \quad A^T = \begin{bmatrix} 1 & 0 & -1 \\ 3 & 2 & 5 \\ -1 & 4 & 7 \end{bmatrix}$$

$$I_2 = \begin{bmatrix} 1 & 0 \\ 0 & 1 \end{bmatrix} \quad I_2^T = \begin{bmatrix} 1 & 0 \\ 0 & 1 \end{bmatrix}$$

i.e. the identity matrix and its transpose are the same.

$$I_3 = \begin{bmatrix} 1 & 0 & 0 \\ 0 & 1 & 0 \\ 0 & 0 & 1 \end{bmatrix} = I_3^T$$

4.2 Origins of Matrices

A matrix defines a linear transformation between vector spaces, namely object vector space with basis x, y, z and image vector space with basis x', y', z'. The transformation due to multiplication by matrix A is

$$\begin{bmatrix} x' \\ y' \\ z' \end{bmatrix} = \begin{bmatrix} a_{11} & a_{12} & a_{13} \\ a_{21} & a_{22} & a_{23} \\ a_{31} & a_{32} & a_{33} \end{bmatrix} \begin{bmatrix} x \\ y \\ z \end{bmatrix}$$

Example: For 3-D vector spaces,

$$\begin{bmatrix} x' \\ y' \\ z' \end{bmatrix} = \begin{bmatrix} 1 & 2 & 3 \\ 0 & 1 & 2 \\ 3 & 0 & 1 \end{bmatrix} \begin{bmatrix} x \\ y \\ z \end{bmatrix}$$

$$\begin{aligned} x' &= x &+ 2y &+ 3z \\ y' &= & y &+ 2z \\ z' &= 3x & &+ z \end{aligned}$$

Example: For 2-D vector spaces,

$$\begin{bmatrix} x' \\ y' \end{bmatrix} = \begin{bmatrix} 1 & 2 \\ 2 & 2 \end{bmatrix} \begin{bmatrix} x \\ y \end{bmatrix} = \begin{bmatrix} x + 2y \\ 2x + 2y \end{bmatrix}$$

Functions in the object plane are transformed into different functions in the image plane (in this case the vector spaces are planes $x'y'$ and xy)
For the example above,

$$\begin{aligned} x' &= x + 2y \\ y' &= 2x + 2y \end{aligned}$$

A straight line $y = x$ in the object plane is transformed into a different straight line in the image plane. The transformed function must also be a straight line since the transformation due to matrix multiplication is always linear (▶Fig. 4.1).
In this case $x' = 3x$, $y' = 4x$, so that the function $y = x$ is transformed into $y' = \frac{4}{3}x'$.
Example:

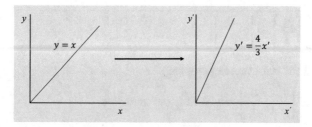

◘ Fig. 4.1 Straight line rotation by matrix multiplication

Transformation of type $\begin{bmatrix} \lambda & 0 \\ 0 & \lambda \end{bmatrix}$ such as $\begin{bmatrix} 4 & 0 \\ 0 & 4 \end{bmatrix} = 4 \begin{bmatrix} 1 & 0 \\ 0 & 1 \end{bmatrix}$

What does the straight line $y = 2x + 3$ transform to in this case?

4

$$\begin{bmatrix} x' \\ y' \end{bmatrix} = 4 \begin{bmatrix} x \\ y \end{bmatrix}$$

$y = 2x + 3 \quad x = \dfrac{x'}{4} \quad y = \dfrac{y'}{4}$

$\dfrac{y'}{4} = 2\dfrac{x'}{4} + 3 \Rightarrow y' = 2x' + 12$

Linear transformations maintain the straight line functionality, but the slope of the line can change, along with the intercepts on the axes, as in the above case.
Example:

$$\begin{bmatrix} x' \\ y' \end{bmatrix} = \begin{bmatrix} 1 & 2 \\ 2 & 2 \end{bmatrix} \begin{bmatrix} x \\ y \end{bmatrix} = \begin{bmatrix} x + 2y \\ 2x + 2y \end{bmatrix}$$

Consider the straight line of constant value $y = 4$

$$\begin{aligned} x' &= x + 2y &= x + 8 \\ y' &= 2x + 2y &= 2x + 8 \\ \therefore y' &= 2x' - 8 \end{aligned}$$

The line $y = 4$ parallel to the x-axis is transformed into the line of slope $m = 2$ with intercept -8 on the y'-axis.

4.2.1 Some More 2D Matrix Operations

Reflection about the x-axis

$$\begin{bmatrix} 1 & 0 \\ 0 & -1 \end{bmatrix} \begin{bmatrix} 0 \\ 1 \end{bmatrix} = \begin{bmatrix} 0 \\ -1 \end{bmatrix}$$

Reflection about the y-axis

$$\begin{bmatrix} -1 & 0 \\ 0 & 1 \end{bmatrix} \begin{bmatrix} 2 \\ 2 \end{bmatrix} = \begin{bmatrix} -2 \\ 2 \end{bmatrix}$$

(See ▶ Fig. 4.2 for both these examples.)

$\begin{bmatrix} 1 & 0 \\ 0 & -1 \end{bmatrix}$ and $\begin{bmatrix} -1 & 0 \\ 0 & 1 \end{bmatrix}$ are "reflection" matrices.

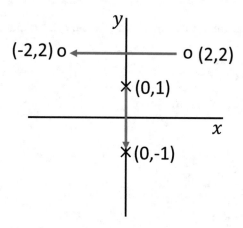

4.3 Inverse of a Matrix A^{-1}

The first thing to note is that, just like for scalars, the identity matrix is its own inverse ($I = I^{-1}$. cf $1^{-1} = \frac{1}{1} = 1$ for scalars)

i.e. $I = \begin{bmatrix} 1 & 0 \\ 0 & 1 \end{bmatrix} = I^{-1}$

$I^{-1}I = \begin{bmatrix} 1 & 0 \\ 0 & 1 \end{bmatrix} \times \begin{bmatrix} 1 & 0 \\ 0 & 1 \end{bmatrix} = \begin{bmatrix} 1 & 0 \\ 0 & 1 \end{bmatrix}$

Before dealing with the inverse for other matrices, we must know about the determinant, minor and cofactors.

The determinant is only defined for a square matrix. e.g. for a 2×2 matrix $A = \begin{bmatrix} a & b \\ c & d \end{bmatrix}$, the determinant is written in two different notations:.

$$\det A = |A| = ad - bc \equiv \begin{vmatrix} a & b \\ c & d \end{vmatrix}$$

Example:

$A = \begin{bmatrix} 3 & 6 \\ 1 & -1 \end{bmatrix}$

$|A| = 3(-1) - 1 \cdot 6 = -9$

The determinant of a matrix is just a number (a scalar). It can be positive or negative depending on the elements of the matrix. For example,

$A = \begin{bmatrix} 8 & 3 \\ -2 & 4 \end{bmatrix}$ $\quad |A| = 38.$

Calculating the determinant of a 3 × 3 matrix is more complicated and is demonstrated in ▶Sect. 4.3.3.

More about determinants

The det operator is distributive.

$$\det(A \times B) = \det A \times \det B$$

Example:

$$A = \begin{bmatrix} 2 & 3 \\ 4 & 1 \end{bmatrix} \quad B = \begin{bmatrix} -3 & 1 \\ 2 & 0 \end{bmatrix}$$

$$A \times B = \begin{bmatrix} 0 & 2 \\ -10 & 4 \end{bmatrix}$$

$$|A \times B| = 20 \quad |A| = -10 \quad |B| = -2$$

$$|A \times B| = 20 = |A| \times |B| = -10 \times -2 = 20$$

If 2 rows or columns are identical $|A| = 0$.
Example:

$$\begin{vmatrix} 2 & 3 \\ 2 & 3 \end{vmatrix} = 0$$

$\det A = \det(A^T)$

Proof for a 2 × 2 matrix:

$$A = \begin{bmatrix} a & b \\ c & d \end{bmatrix} \quad A^T = \begin{bmatrix} a & c \\ b & d \end{bmatrix}$$

$$|A| = ad - bc = |A^T|$$

4.3.1 Minor of a Matrix Element

This is key to calculating the determinant of a 3 × 3 matrix and larger.
Example:

$$A = \begin{bmatrix} 3 & 9 & 0 \\ -5 & -1 & 3 \\ 4 & 7 & 0 \end{bmatrix}$$

Strike out the row and column through an element a_{ij}. The determinant of the matrix formed is the minor of a_{ij}. For example,

$$M_{12} = \begin{vmatrix} -5 & 3 \\ 4 & 0 \end{vmatrix} = -12$$

$$M_{22} = \begin{vmatrix} 3 & 0 \\ 4 & 0 \end{vmatrix} = 0$$

The same principle can be applied to matrices of any rank, including 2×2 matrices. Thus,

$$B = \begin{bmatrix} 1 & 2 \\ 3 & 1 \end{bmatrix}$$

$M_{11} = 1;\ M_{12} = 3$

$M_{21} = 2;\ M_{22} = 1$

4.3.2 Cofactor of a Matrix Element

The cofactor is numerically equal to the minor but with a sign which depends on the position of the element.

$$C_{ij} = (-1)^{i+j} M_{ij}$$

The term $(-1)^{i+j}$ is $+1$ or -1 depending on the position of the element a_{ij} in the matrix and is often called the place sign.

Example: In a 3×3 matrix the place signs are $\begin{bmatrix} + & - & + \\ - & + & - \\ + & - & + \end{bmatrix}$. For A, above,

$C_{23} = (-1) \times -15 = 15.$

4.3.3 An Application of Minors and Cofactors

Example: Determinant of a 3×3 matrix

$$A = \begin{bmatrix} a_{11} & a_{12} & a_{13} \\ a_{21} & a_{22} & a_{23} \\ a_{31} & a_{32} & a_{33} \end{bmatrix}$$

Expanding along row 1 gives the determinant in terms of the cofactors of the 1st row elements

$$|A| = a_{11}C_{11} + a_{12}C_{12} + a_{13}C_{13}$$

Remember the signs in the cofactors

$$C_{ij} = (-1)^{i+j} M_{ij}$$

Example:

$$A = \begin{bmatrix} 3 & -1 & 6 \\ 9 & -5 & 2 \\ 0 & 4 & 7 \end{bmatrix}$$

Expanding along row 1,

$$|A| = 3\begin{vmatrix} -5 & 2 \\ 4 & 7 \end{vmatrix} - (-1)\begin{vmatrix} 9 & 2 \\ 0 & 7 \end{vmatrix} + 6\begin{vmatrix} 9 & -5 \\ 0 & 4 \end{vmatrix}$$

$$= 3(-43) + 9(7) + 6(36)$$

$$= -129 + 63 + 216 = 150$$

Simplifying |A|

Since $|A|$ is independent of the row chosen for expansion, calculation of $|A|$ is simplified by choosing a row containing zeroes.

Example:

$$A = \begin{bmatrix} 1 & 2 & 5 \\ 3 & 1 & 2 \\ 0 & 0 & 3 \end{bmatrix}$$

From row 1, $|A| = 1(3) - 2(3 \times 3) + 5 \times 0 = -15$

From row 3, $|A| = 3[1(1) - 3(2)] = -15$

The same result is obtained more easily by expanding from row 3 which contains two zero elements.

4.3.4 Adjugate or Adjoint Matrix

The adjugate/adjoint of a square matrix is the transpose of its cofactor matrix (matrix of cofactors).

$$\text{adj}(A) = C^T$$

The inverse matrix of A is then constructed from the adjoint:

$$A^{-1} = \frac{1}{|A|}\text{adj}(A)$$

Example:

$$A = \begin{bmatrix} 3 & 4 \\ 2 & 5 \end{bmatrix} \qquad C = \begin{bmatrix} 5 & -2 \\ -4 & 3 \end{bmatrix}$$

$$\text{adj}(A) = C^T = \begin{bmatrix} 5 & -4 \\ -2 & 3 \end{bmatrix}$$

$$|A| = 15 - 8 = 7$$

$$\therefore A^{-1} = \frac{1}{7}\begin{bmatrix} 5 & -4 \\ -2 & 3 \end{bmatrix}$$

Check:

$$A^{-1}A = \frac{1}{7}\begin{bmatrix} 5 & -4 \\ -2 & 3 \end{bmatrix}\begin{bmatrix} 3 & 4 \\ 2 & 5 \end{bmatrix} = \frac{1}{7}\begin{bmatrix} 7 & 0 \\ 0 & 7 \end{bmatrix}$$

$$\therefore AA^{-1} = \begin{bmatrix} 1 & 0 \\ 0 & 1 \end{bmatrix} = I_2 \text{ as required}$$

4.4 Inverse of a 3 × 3 Matrix

$$A = \begin{bmatrix} 1 & 1 & 2 \\ 2 & 1 & 1 \\ 0 & 0 & 2 \end{bmatrix}$$

$$\begin{aligned}
|A| &= 1\begin{vmatrix} 1 & 1 \\ 0 & 2 \end{vmatrix} - 1\begin{vmatrix} 2 & 1 \\ 0 & 2 \end{vmatrix} + 2\begin{vmatrix} 2 & 1 \\ 0 & 0 \end{vmatrix} \\
&= 2 - 4 + 0 = -2
\end{aligned}$$

$$C_{ij} = (-1)^{i+j}M_{ij}$$

$$C_{11} = \begin{vmatrix} 1 & 1 \\ 0 & 2 \end{vmatrix} \quad C_{12} = -\begin{vmatrix} 2 & 1 \\ 0 & 2 \end{vmatrix} \quad C_{13} = \begin{vmatrix} 2 & 1 \\ 0 & 0 \end{vmatrix}$$

$$C_{21} = -\begin{vmatrix} 1 & 2 \\ 0 & 2 \end{vmatrix} \quad C_{22} = \begin{vmatrix} 1 & 2 \\ 0 & 2 \end{vmatrix} \quad C_{23} = -\begin{vmatrix} 1 & 1 \\ 0 & 0 \end{vmatrix}$$

$$C_{31} = \begin{vmatrix} 1 & 2 \\ 1 & 1 \end{vmatrix} \quad C_{32} = -\begin{vmatrix} 1 & 2 \\ 2 & 1 \end{vmatrix} \quad C_{33} = \begin{vmatrix} 1 & 1 \\ 2 & 1 \end{vmatrix}$$

$$C = \begin{bmatrix} 2 & -4 & 0 \\ -2 & 2 & 0 \\ -1 & 3 & -1 \end{bmatrix} \quad C_T = \begin{bmatrix} 2 & -2 & -1 \\ -4 & 2 & 3 \\ 0 & 0 & -1 \end{bmatrix}$$

$$A^{-1} = \frac{1}{|A|}\text{adj}A = -\frac{1}{2}\begin{bmatrix} 2 & -2 & -1 \\ -4 & 2 & 3 \\ 0 & 0 & -1 \end{bmatrix} = \begin{bmatrix} -1 & 1 & \frac{1}{2} \\ 2 & -1 & -\frac{3}{2} \\ 0 & 0 & \frac{1}{2} \end{bmatrix}$$

Check:

$$\begin{bmatrix} -1 & 1 & \frac{1}{2} \\ 2 & -1 & -\frac{3}{2} \\ 0 & 0 & \frac{1}{2} \end{bmatrix}\begin{bmatrix} 1 & 1 & 2 \\ 2 & 1 & 1 \\ 0 & 0 & 2 \end{bmatrix} = \begin{bmatrix} 1 & 0 & 0 \\ 0 & 1 & 0 \\ 0 & 0 & 1 \end{bmatrix}$$

If $|A| = 0$, it is clear that A^{-1} is not defined. Calculating $|A|$ first gives a quick check that $\exists A^{-1}$. Unless $|A| \neq 0$ there is no need to determine $\mathrm{adj}A$: the matrix has no inverse.

4.5 Matrices for Simultaneous Equations

Consider a simple system of linear simultaneous equations:

$$x + y = 3$$
$$3x + 2y = 5$$

In matrix form

$$\begin{bmatrix} 1 & 1 \\ 3 & 2 \end{bmatrix} \begin{bmatrix} x \\ y \end{bmatrix} = \begin{bmatrix} 3 \\ 5 \end{bmatrix}$$

$|A| = -1 \Rightarrow \exists A^{-1}$ since $A| \neq 0$.

$$A^{-1} = -\begin{bmatrix} 2 & -3 \\ -1 & 1 \end{bmatrix}^T = \begin{bmatrix} -2 & 1 \\ 3 & -1 \end{bmatrix}$$

$$\begin{bmatrix} -2 & 1 \\ 3 & -1 \end{bmatrix} \begin{bmatrix} 1 & 1 \\ 3 & 2 \end{bmatrix} \begin{bmatrix} x \\ y \end{bmatrix} = \begin{bmatrix} 1 & 0 \\ 0 & 1 \end{bmatrix} \begin{bmatrix} x \\ y \end{bmatrix}$$

$$= \begin{bmatrix} -2 & 1 \\ 3 & -1 \end{bmatrix} \begin{bmatrix} 3 \\ 5 \end{bmatrix}$$

$$\therefore \begin{bmatrix} x \\ y \end{bmatrix} = \begin{bmatrix} -1 \\ 4 \end{bmatrix}$$

The method appears complicated, but is useful as a general method for solving larger systems of linear simultaneous equations.

Example for 3 unknowns:

$$2x + y - 2z = 4 \quad (i)$$
$$2x + 3y + z = 2 \quad ii$$
$$x \qquad - 2z = 3 \quad (iii)$$

Method 1: Manipulation of Equations

(i)	$\Rightarrow$	$6x$	$+$	$3y$	$-$	$6z$	$= 12$	(iv)
$(iv), (ii)$	$\Rightarrow$			$4x$	$-$	$7z$	$= 10$	(v)
(iii)	$\Rightarrow$			$4x$	$-$	$8z$	$= 12$	(vi)
$(v), (vi)$	$\Rightarrow$	z	$=$	-2				
From (iii),		x	$=$	-1				
From (iv),		y	$=$	2				
Solution:	x	$= -1,$	y	$= 2,$	z	$= -2$		

Method 2: Matrix Algebra

$$\begin{bmatrix} 2 & 1 & -2 \\ 2 & 3 & 1 \\ 1 & 0 & -2 \end{bmatrix} \begin{bmatrix} x \\ y \\ z \end{bmatrix} = \begin{bmatrix} 4 \\ 2 \\ 3 \end{bmatrix}$$

Check $\exists$ a solution, expanding from row 3 for simplicity

$$|A| = \begin{vmatrix} 1 & -2 \\ 3 & 1 \end{vmatrix} - 2 \begin{vmatrix} 2 & 1 \\ 2 & 3 \end{vmatrix} = 7 - 2 \times 4 = -1 \neq 0$$

Thus, $\exists \, A^{-1}$ and a solution to the system of equations. Calculating the matrix of cofactors,

$$A^{-1} = \frac{1}{|A|} \text{adj} A = - \begin{bmatrix} -6 & 5 & -3 \\ 2 & -2 & 1 \\ 7 & -6 & 4 \end{bmatrix}^{T}$$

$$= \begin{bmatrix} 6 & -2 & -7 \\ -5 & 2 & 6 \\ 3 & -1 & -4 \end{bmatrix}$$

$$\therefore \begin{bmatrix} x \\ y \\ z \end{bmatrix} = \begin{bmatrix} 6 & -2 & -7 \\ -5 & 2 & 6 \\ 3 & -1 & -4 \end{bmatrix} \begin{bmatrix} 4 \\ 2 \\ 3 \end{bmatrix} = \begin{bmatrix} -1 \\ 2 \\ -2 \end{bmatrix}$$

The identical result to Method 1.

Method 3: Gaussian Elimination
This method involves reducing an augmented matrix representing the system of equations to echelon form or as near to echelon form as possible, given the numbers involved. The methodology resembles Method 1.
A single non square "augmented" matrix is formed from both sides of the system of equations.

$$\begin{bmatrix} 2 & 1 & -2 & \vdots & 4 \\ 2 & 3 & 1 & \vdots & 2 \\ 1 & 0 & -2 & \vdots & 3 \end{bmatrix}$$

Reduce this to echelon form by combining rows, switching rows, etc.

$$\xrightarrow{R1-R2} \begin{bmatrix} 0 & -2 & -3 & | & 2 \\ 2 & 3 & 1 & | & 2 \\ 1 & 0 & -2 & | & 3 \end{bmatrix} \xrightarrow{R2-2R3} \begin{bmatrix} 0 & -2 & -2 & | & -2 \\ 0 & 3 & 5 & | & -4 \\ 1 & 0 & -2 & | & -3 \end{bmatrix}$$

$$\xrightarrow{\text{Interchange } R1, R3} \begin{bmatrix} 1 & 0 & -2 & | & 3 \\ 0 & 3 & 5 & | & -4 \\ 0 & -2 & -3 & | & 2 \end{bmatrix}$$

$$\xrightarrow{\frac{3}{2} \times R3} \begin{bmatrix} 1 & 0 & -2 & | & 3 \\ 0 & 3 & 5 & | & -4 \\ 0 & -3 & \frac{-9}{2} & | & 3 \end{bmatrix}$$

$$\xrightarrow{R3+R2} \begin{bmatrix} 1 & 0 & -2 & | & 3 \\ 0 & 3 & 5 & | & -4 \\ 0 & 0 & \frac{1}{2} & | & -1 \end{bmatrix}$$

The part of the matrix formed from the coefficients has now been diagonalizd and the overall augmented matrix is in row echelon form, from which the solution can be read as

$$\text{Row3} \Rightarrow \frac{z}{2} = -1 \Rightarrow z = -2$$
$$\text{Row2} \Rightarrow 3y + 5z = -4 \Rightarrow y = 2$$
$$\text{Row1} \Rightarrow x + 0(-2) - 2(-2) = 3$$
$$\Rightarrow x = -1$$

4.6 Eigenvalues and Eigenvectors

These arise when a matrix equation takes a particular form. The standard matrix form representing a system of linear equations is, for the case of 3 variables,

$$\begin{bmatrix} a_{11} & a_{12} & a_{13} \\ a_{21} & a_{22} & a_{23} \\ a_{31} & a_{32} & a_{33} \end{bmatrix} \begin{bmatrix} x_1 \\ x_2 \\ x_3 \end{bmatrix} = \begin{bmatrix} b_1 \\ b_2 \\ b_3 \end{bmatrix}$$

from which, as shown in the previous section, the unknown vector $\begin{bmatrix} x_1 \\ x_2 \\ x_3 \end{bmatrix}$ can be determined. A special case of this problem in which the underlying equations describe a system in engineering or physics is

$$\begin{bmatrix} a_{11} & a_{12} & a_{13} \\ a_{21} & a_{22} & a_{23} \\ a_{31} & a_{32} & a_{33} \end{bmatrix} \begin{bmatrix} x_1 \\ x_2 \\ x_3 \end{bmatrix} = \lambda \begin{bmatrix} x_1 \\ x_2 \\ x_3 \end{bmatrix}$$

In operator form:

$$AX_n = \lambda_n X_n$$

Operator equations of this type occur in quantum mechanics and classical mechanics, for example. The matrix operator A acts on vector X with the effect of multiplying X by a scalar λ. In general, $\exists\, n$ independent solutions in the form of pairs of scalar λ_n and vector X_n satisfying this operator equation. λ_n are called the eigenvalues of the problem and X_n are the corresponding eigenvectors for which $\exists$ a one-to-one relationship

$$\lambda_n \leftrightarrow X_n$$

The "homogeneous" form of the eigenvalue problem with zero vector on the rhs is obtained by rearranging to give

$$\begin{bmatrix} a_{11} - \lambda & a_{12} & a_{13} \\ a_{21} & a_{22} - \lambda & a_{23} \\ a_{31} & a_{32} & a_{33} - \lambda \end{bmatrix} \begin{bmatrix} x_1 \\ x_2 \\ x_3 \end{bmatrix} = \begin{bmatrix} 0 \\ 0 \\ 0 \end{bmatrix}$$

Example:

$$\begin{bmatrix} 2 & 3 \\ 4 & 1 \end{bmatrix} \begin{bmatrix} x_1 \\ x_2 \end{bmatrix} = \lambda \begin{bmatrix} x_1 \\ x_2 \end{bmatrix}$$

$$\begin{bmatrix} 2 - \lambda & 3 \\ 4 & 1 - \lambda \end{bmatrix} \begin{bmatrix} x_1 \\ x_2 \end{bmatrix} = \begin{bmatrix} 0 \\ 0 \end{bmatrix}$$

or $\quad (A - \lambda I_2) = 0$

$\therefore |A - \lambda I_2||X| = 0$

The solution is trivial and $X = \begin{bmatrix} 0 \\ 0 \end{bmatrix}$ except in the case that $|A - \lambda I_2| = 0$.

i.e. $\quad \begin{vmatrix} 2 - \lambda & 3 \\ 4 & 1 - \lambda \end{vmatrix} = (2 - \lambda)(1 - \lambda) - 12 = 0$

$\therefore \lambda^2 - 3\lambda - 10 = 0$

$(\lambda - 5)(\lambda + 2) = 0$

$\Rightarrow \lambda_1 = 5, \lambda_2 = -2$

The problem has 2 characteristic roots or eigenvalues, λ_1 and λ_2.
Knowing the eigenvalues $\lambda_1 = 5, \lambda_2 = -2$, the corresponding eigenvectors X_1, X_2 can be determined from the homogeneous equation.

$$\begin{bmatrix} 2 - \lambda & 3 \\ 4 & 1 - \lambda \end{bmatrix} \begin{bmatrix} x_1 \\ x_2 \end{bmatrix} = \begin{bmatrix} 0 \\ 0 \end{bmatrix}$$

$\lambda = 5: \quad \begin{bmatrix} -3 & 3 \\ 4 & -4 \end{bmatrix} \begin{bmatrix} x_1 \\ x_2 \end{bmatrix} = \begin{bmatrix} 0 \\ 0 \end{bmatrix}$

so that $x_1 = x_2 = x$ and the first eigenvector is

$$X_1 = \begin{bmatrix} x \\ x \end{bmatrix} = x \begin{bmatrix} 1 \\ 1 \end{bmatrix}$$

We have free choice of x, so that, choosing $x = 1$, the first eigenvector is $\begin{bmatrix} 1 \\ 1 \end{bmatrix}$ and any multiple of this, integer or non-integer, is also a first eigenvector. For the second eigenvalue:

$$\lambda = -2: \quad \begin{bmatrix} 4 & 3 \\ 4 & 3 \end{bmatrix} \begin{bmatrix} x_1 \\ x_2 \end{bmatrix} \begin{bmatrix} 0 \\ 0 \end{bmatrix}$$

so that $x_2 = -\frac{4}{3}x_1$ and the second eigenvector is

$$X_2 = \begin{bmatrix} x \\ -\frac{4}{3}x \end{bmatrix} = x \begin{bmatrix} 1 \\ -\frac{4}{3} \end{bmatrix} = \frac{x}{3} \begin{bmatrix} 3 \\ -4 \end{bmatrix}$$

again with free choice of x.

$$x = 1 \Rightarrow X_2 = \begin{bmatrix} 3 \\ -4 \end{bmatrix} \quad \text{Check:} \quad \begin{bmatrix} 4 & 3 \\ 4 & 3 \end{bmatrix} \begin{bmatrix} 3 \\ -4 \end{bmatrix} = \begin{bmatrix} 0 \\ 0 \end{bmatrix}$$

Any scalar multiple is also an eigenvector (x does not have to take an integer value).

e.g. $x = \dfrac{1}{3} \Rightarrow X_2 = \begin{bmatrix} 1 \\ -\frac{4}{3} \end{bmatrix}$

$$\begin{bmatrix} 4 & 3 \\ 4 & 3 \end{bmatrix} \begin{bmatrix} 1 \\ -\frac{4}{3} \end{bmatrix} = \begin{bmatrix} 0 \\ 0 \end{bmatrix}$$

4.7 Matrices in System Stability

Matrices are also relevant to engineering control theory when the State Matrix of a system is used to determine the so-called poles and to reveal whether the system is stable. (This topic is also considered in Chap. 6, Sect. 6.6 in the context of partial fractions in integration.)

Consider a system in the form of a damped mechanical oscillator, as in Fig. 10.1, Sect. 10.1. The natural oscillations are described by the 2nd order ODE.[1]

$$\ddot{y} + b\dot{y}(t) + \omega^2 y(t) = 0$$

---

[1]   See Chapter 10 for a more complete discussion of damped oscillations.

where b is the velocity dependent damping coefficient and
$\omega = \sqrt{\dfrac{k}{m}}$ is the natural oscillation frequency. If the system is in forced oscillation, the
equation of motion is modified to include, on the rhs, a "forcing term" $u(t) = \dfrac{F(t)}{m}$,
where $F(t)$ is the time varying applied force. For example, this might be an impulsive
force applied at $t = 0$, or a continuous sinusoidal force of form $\sin pt$ where, in
general, $p \neq \omega$.

The modified equation of motion is

$$\ddot{y} + b\dot{y}(t) + \omega^2 y(t) = u(t)$$

This equation together with its boundary conditions describes the mechanical oscillator system and may be converted to matrix form by introducing the state vector
$\begin{bmatrix} x_1(t) \\ x_2(t) \end{bmatrix}$ where $x_1(t)$ is position $y(t)$, $x_2(t)$ is velocity $\dot{y}(t)$.

Note, $\dot{x}_1(t) = x_2(t); \; \dot{x}_2(t) = \ddot{y}(t)$ so that the above equation becomes:

$$\begin{bmatrix} \dot{x}_1 \\ \dot{x}_2 \end{bmatrix} = \begin{bmatrix} 0 & 1 \\ -\omega^2 & -b \end{bmatrix} \begin{bmatrix} x_1 \\ x_2 \end{bmatrix} + \begin{bmatrix} 0 \\ 1 \end{bmatrix} u(t)$$

This is the "State Equation" and the matrix $A = \begin{bmatrix} 0 & 1 \\ -\omega^2 & -b \end{bmatrix}$ is the "State
Matrix".

Example: A damped mechanical oscillator is described by the equation of motion

$$\ddot{y} + 4\dot{y} + 3y = 2\sin 3t$$

In matrix form,

$$\begin{bmatrix} \dot{x}_1 \\ \dot{x}_2 \end{bmatrix} = \begin{bmatrix} 0 & 1 \\ -3 & -4 \end{bmatrix} \begin{bmatrix} x_1 \\ x_2 \end{bmatrix} + 2\sin 3t \begin{bmatrix} 0 \\ 1 \end{bmatrix}$$

The term $\sin 3t \begin{bmatrix} 0 \\ 1 \end{bmatrix}$ is the input to the system; the system response is determined
by the State Matrix A, from which the stability of the system is indicated by the
poles in the s-domain (see Sect. 6.6) which are found from

$$|A - sI| = 0$$

$$\therefore \; \begin{vmatrix} -s & 1 \\ -3 & -4 - s \end{vmatrix} = 0$$

$$\Rightarrow s^2 + 4s + 3 = 0$$

$$(s + 3)(s + 1) = 0$$

$$s = -1, -3$$

The poles are in the left half plane which, in Control Theory, is the condition that the system is stable. [See Sect. 6.6, page 58 for a limited explanation of the science and mathematics of Control Theory using the method of Laplace Transforms.]

4.7.1 Note on $|A - sI| = 0$

The elements in the matrix of poles may be complex numbers.
Example:
Find the system poles if the State Matrix is $\begin{bmatrix} -1 & 1 \\ -5 & 2 \end{bmatrix}$.

$$\begin{vmatrix} -(1+s) & 1 \\ -5 & 2-s \end{vmatrix} = 0$$

$$\Rightarrow s^2 - s + 3 = 0$$

$$\therefore s = \frac{1}{2} \pm \frac{i\sqrt{11}}{2}$$

Both poles lie on the rhs of $\mathbb{C}$, so that the system is unstable.

Differentiation

Contents

The derivative of a function $y(x)$ is a way of determining its slope at any point P in the xy−plane, written $\dfrac{dy}{dx}$; see ◻ Fig. 5.1 from which

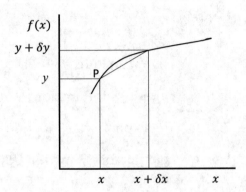

◻ **Fig. 5.1** Construction of the derivative

$$\frac{dy}{dx} = \lim_{\delta x \to 0} \frac{\delta y}{\delta x}$$

This is the slope of the curve $y(x)$ at P.
For the straight line $y = x$,

$$\frac{dy}{dx} = \lim_{\delta x \to 0} \frac{(x + \delta x - x)}{\delta x}$$
$$= 1$$

If $y = x^2$

$$\frac{dy}{dx} = \lim_{\delta x \to 0} \left[\frac{(x + \delta x)^2 - x^2}{\delta x} \right]$$
$$= \lim_{\delta x \to 0} \left[\frac{2x\delta x + \delta x^2}{\delta x} \right]$$
$$= 2x$$

If $y = x^3$

$$\frac{dy}{dx} = \lim_{\delta x \to 0} \left[\frac{x^3 + 3x^2\delta x + 3x\delta x^2 + \delta x^3 - x^3}{\delta x} \right]$$
$$= 3x^2$$

Exercise: Show that $\dfrac{d}{dx}(x^4) = 4x^3$

By induction $\dfrac{d}{dx}(x^n) = nx^{n-1}$.

If $y = $ constant c

$$\frac{dy}{dx} = 0$$

If $y = af(x)$

$$\frac{dy}{dx} = a\frac{df}{dx}$$

e.g.

$$\begin{aligned} y &= 5x^3 \\ \frac{dy}{dx} &= 5 \cdot 3x^2 = 15x^2 \end{aligned}$$

This can be understood through the Chain Rule, as follows.

5.1 The Chain Rule

$$\frac{d}{dx}(u(x)v(x)) = u\frac{dv}{dx} + v\frac{du}{dx}$$

Proof:

$$\begin{aligned} \frac{d}{dx}(uv) &= \lim_{\delta x \to 0} \frac{\delta(uv)}{\delta x} \\ &= \lim_{\delta x \to 0} \left[\frac{u(x+\delta x)v(x+\delta x) - u(x)v(x)}{\delta x} \right] \\ &= \lim_{\delta x \to 0} \left[\frac{(u+\delta u)(v+\delta v) - uv}{\delta x} \right] \\ &= \lim_{\delta x \to 0} \left[u\frac{\delta v}{\delta x} + v\frac{\delta u}{\delta x} + \frac{\delta u \delta v}{\delta x} \right] \end{aligned}$$

The 3rd term is zero in the limit as $\delta x \to 0$, therefore

$$\frac{d(uv)}{dx} = u\frac{dv}{dx} + v\frac{du}{dx}$$

Exercise: Show that $\dfrac{d}{dt}\left(\dfrac{u}{v}\right) = \dfrac{v\frac{du}{dx} - u\frac{dv}{dx}}{v^2}$

5.2 Logarithmic Functions

$$
\begin{aligned}
y &= a^x & \log_a y &= x \\
y &= 10^x & \log_{10} y &= x \\
y &= e^x & \log_e y &= \ln y = x
\end{aligned}
$$

$$
\frac{d}{dx}(\ln x) = \frac{1}{x}
$$

5

Proof:

$$
\begin{aligned}
y &= \ln x \\
x &= e^{\ln x} = e^y \\
\frac{dx}{dy} &= e^y = x \\
\frac{dy}{dx} &= \frac{1}{x}
\end{aligned}
$$

$$
\frac{d}{dx}(\ln x) = \frac{1}{x}
$$

5.3 The Exponential Function

A unique function which is identical to its derivative:

$$
\begin{aligned}
y(x) &= e^x = \exp x \\
\frac{dy}{dx} &= e^x
\end{aligned}
$$

Writing e^x as an infinite series,

$$
\begin{aligned}
y(x) &= e^x = 1 + x + \frac{x^2}{2!} + \frac{x^3}{3!} + \cdots + \frac{x^n}{n!} + \cdots \\
\frac{dy}{dx} &= 0 + 1 + x + \frac{x^2}{2!} + \cdots + \frac{x^n}{n!} + \cdots \\
&= e^x
\end{aligned}
$$

5.4 Function of a Function

$$
y = f(g(x))
$$

$$
\frac{dy}{dx} = \frac{df}{dg}\frac{dg}{dx}
$$

e.g. $y = e^{ax}$ $y = f(g(x))$ where $f = e^g$, $g = ax$

$$\frac{dy}{dx} = e^g \frac{dg}{dx} = e^{ax} a$$

$$\frac{dy}{dx} = ae^{ax}$$

Some things we take for granted in differential calculus:

$$\frac{d}{dx}(f(x) + g(x)) = \frac{df}{dx} + \frac{dg}{dx}$$
$$= \frac{dg}{dx} + \frac{df}{dx}$$

It doesn't matter which way round we form the sum because, for real numbers,

$$a + b = b + a$$

(commutative)

Integration

Contents

© The Author(s), under exclusive license to Springer Nature Switzerland AG 2022
P. Prewett, *Foundation Mathematics for Science and Engineering Students*,
https://doi.org/10.1007/978-3-030-91963-4_6

6.1 **The Meaning of Integration**

Integration is often treated as anti-differentiation, in which the integral of a function is what must be differentiated to get the function back. But $I = \int y \mathrm{d}x$ is also the area under the curve $y(x)$. The segmented area $= \sum_{x=a}^{b} f(x)\delta x$ in ►Fig. 6.1, from which the definite integral of $f(x)$ on the interval $a < x < b$ is defined as[1]

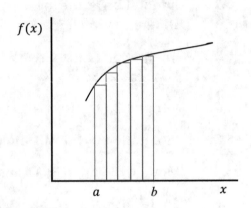

□ **Fig. 6.1** The basis of integration

$$I = \lim_{\delta x \to 0} \sum_{a}^{b} f(x)\delta x = \int_{a}^{b} f(x)\mathrm{d}x$$

The indefinite integral, written $\int y(x)\mathrm{d}x$ is a function of x whereas the definite integral, summed over a specific interval $[a, b]$ takes a numerical value (written $\int_{a}^{b} f(x)\mathrm{d}x$). The function to be integrated is called the integrand.

6.2 **Integration as the Inverse of Differentiation**

Powers of x:

$$\begin{aligned} I &= \int x^n \mathrm{d}x \\ &= \frac{x^{n+1}}{n+1} + c \end{aligned}$$

---

1   [Newton and Leibnitz quarrelled over the development of Integral Calculus. The common notation is due to Leibnitz.]

Proof: Let

$$I = \frac{x^{n+1}}{n+1} + c$$

Then

$$\frac{dI}{dx} = x^n$$

$$I = \int \frac{dI}{dx} dx = \int x^n dx = \frac{x^{n+1}}{n+1} + c$$

6.3 Integration of the Circular Functions

(a)

$$I = \int \cos x \, dx$$

$$\cos x = \frac{d}{dx}(\sin x)$$

$$\therefore I = \int \cos x \, dx = \int d(\sin x)$$

$$\therefore I = \sin x + c$$

(b) Exercise:

Show that $\int \sin x \, dx = -\cos x + c$

(c) Indefinite integrals are functions with an arbitrary constant unless we have more information, in which case the constant is defined. This additional information is called a boundary condition.

Thus, if $I = \int f(x) dx$ where $f(x) = \sin x$, then
$I = \int \cos x \, dx = \sin x + c$.
If we know that $I = 0$ when $x = \frac{\pi}{2}$, then c is defined and the integral is

$$I(x) = \sin x - 1$$

(d) A definite integral is always a number, unlike an indefinite integral which is a function of x.

Example:

$$
\begin{aligned}
I &= \int_0^\pi \sin x \, dx &= [-\cos x]_0^\pi \\
&= -(-1) + 1 &= 2
\end{aligned}
$$

(e) Integrating a Function of a Function

e.g. $\displaystyle \int \cos 3x \, dx = \frac{1}{3} \sin 3x + c$

(i) As the inverse of differentiation

6

$$
\frac{d}{dx}(\frac{1}{3} \sin 3x + c) = \frac{1}{3} \times 3 \cos 3x = \cos 3x
$$

(ii) General Proof:

$$
\begin{aligned}
I &= \int f(g(x)) \, dx \\
\therefore I &= \int f(g) dg \frac{dx}{dg} = \int \frac{f(g)}{g'(x)} \, dg.
\end{aligned}
$$

Special case: if g is a *linear* function of x,
i.e. $g = ax + b$, then

$$
\begin{aligned}
I &= \int \frac{f(g)}{a} \, dg \\
\therefore I &= \frac{1}{a} \int f(g) \, dg.
\end{aligned}
$$

If g is not linear in x, the integral cannot be simplified by moving g' outside the $\int$ sign. Other methods of evaluation are then needed.

6.4 Integration by Parts

Consider the definite integral of the form

$$
I = \int_a^b u \frac{dv}{dx} \, dx
$$

This has solution of form

$$
I = [uv]_a^b - \int_a^b v \frac{du}{dx} \, dx
$$

Proof:

$$\frac{d}{dx}(uv) = u\frac{dv}{dx} + v\frac{du}{dx}$$

$$\therefore \int_a^b d(uv) = \int_a^b u\frac{dv}{dx}dx + \int_a^b v\frac{dv}{dx}dx$$

[Incidentally, this step illustrates the use of $\int_a^b dx$ as a distributive operator.]
The proof is concluded by rearranging to give

$$\int_a^b u\frac{dv}{dx}dx = [uv]_a^b - \int_a^b v\frac{dv}{dx}dx$$

Example:

$$\begin{aligned} I &= \int_0^{\frac{\pi}{2}} x\cos x dx \\ &= [x\sin x]_0^{\frac{\pi}{2}} - \int_0^{\frac{\pi}{2}} \sin x dx \\ &= [x\sin x + \cos x]_0^{\frac{\pi}{2}} = \frac{\pi}{2} - 1 \end{aligned}$$

6.5 Integration by Substitution

There are several standard integral forms which are readily solved by substitution.
These are best understood by practice.
For example, $I = \int \frac{dx}{\sqrt{a^2 - x^2}}$. Substituting $x = a\sin\theta$ gives

$$I = \int \frac{a\cos\theta d\theta}{a\sqrt{1 - sin^2\theta}} = \int \frac{\cos\theta d\theta}{\sqrt{1 - \sin^2\theta}}$$

The substitution allows us to use Pythagoras' Theorem $\sin^2\theta + \cos^2\theta = 1$, so that

$$I = \int d\theta = \theta + c$$

Since $x = a\sin\theta$, this solution may be expressed as

$$I(x) = \sin^{-1}\frac{x}{a} + c$$

Exercise:
Show that $\int \frac{dx}{\sqrt{a^2 + x^2}}$ has solution $I(x) = \cos^{-1}\frac{x}{a} + c$.

6.6 Method of Partial Fractions

This method makes use of $\int \dfrac{1}{x} dx = \ln x + c$.

Splitting functions into partial fractions is often used to simplify integration.
Example:

$$I = \int \frac{x+2}{x^2 + 4x + 3} dx$$

Factorising the denominator,

$$\frac{x+2}{x^2 + 4x + 3} = \frac{x+2}{(x+3)(x+1)}$$

$$\equiv \frac{A}{x+3} + \frac{B}{x+1} = \frac{A(x+1) + B(x+3)}{(x+3)(x+1)}$$

Comparing terms;

$$\left. \begin{array}{lll} \varnothing x & : & A + B = 1 \\ \varnothing x^0 & : & A + 3B = 2 \end{array} \right\} \Rightarrow A = B = \frac{1}{2}$$

$$\therefore I = \frac{1}{2} \int \frac{dx}{x+3} + \frac{1}{2} \int \frac{dx}{x+1}$$

$$= \frac{1}{2} [\ln(x+3) + \ln(x+1)] + c$$

$$\therefore I = \frac{1}{2} \ln(x^2 + 4x + 3) + c$$

The method of partial fractions is used in System Control Theory in which the time response of mechanical and electrical systems to impulses is described mathematically using the method of Laplace Transforms. (This is also mentioned briefly in ▶Chap. 4.) In this method, functions are transformed from the time domain to the s-domain using the Integral Transform

$$F(s) = \int_0^\infty e^{-st} f(t) dt$$

The detailed theory of Laplace Transforms is beyond the scope of this text but a short incomplete description of a few representative problems illustrates some of the mathematics involved.

Example: The Laplace Transform $F(s)$ of the time response $f(t)$ of a particular system is

$$F(s) = \frac{3s + 5}{s^2 + 5s + 6}$$

Factorizing the denominator,

$$F(s) = \frac{3s + 5}{(s + 3)(s + 2)} \equiv \frac{A}{s + 3} + \frac{B}{s + 2}$$

$$\left.\begin{array}{rcl} Os \ : & A + B & = & 3 \\ Os^0 \ : & 2A + 3B & = & 5 \end{array}\right\} \Rightarrow A = 4, B = -1$$

$$F(s) = \frac{4}{s + 3} - \frac{1}{s + 2}$$

The solution $F(s)$ in the s-domain is now transformed back to the t-domain, using the fact that the Laplace Transform of e^{-at} is

$$\mathcal{L}(e^{-at}) = \int_0^\infty e^{-at} e^{-st} dt$$

$$= \int_0^\infty e^{-(s+a)t} dt$$

$$= -\left[\frac{e^{-(s+a)t}}{s + a}\right]_{t=0}^\infty$$

$$= \frac{1}{s + a}$$

Hence, the inverse Laplace Transform of $\frac{1}{s + a}$ is

$$\mathcal{L}^{-1}\left\{\frac{1}{s + a}\right\} = e^{-at}.$$

The time response of the system is, therefore

$$f(t) = \mathcal{L}^{-1}\left\{\frac{4}{s + 3} - \frac{1}{s + 2}\right\},$$

so that $f(t) = 4e^{-3t} - e^{-2t}$

In control theory, this represents a stable system since $f(t) \rightarrow 3$ when $t \rightarrow \infty$. The system decays to stable equilibrium after a long time.

Zeroes and Poles: The solutions of $F(s)$ for which $F = 0$ are called the zeroes; the values of s for which $F \rightarrow \infty$ are its poles.

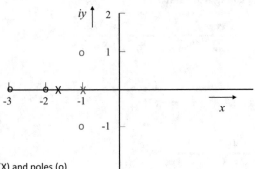

Problem 1 zeroes (X) and poles (o)

Problem 2 zeroes (X) and poles (o)

6

☐ **Fig. 6.2**   Zeroes and poles in the complex plane

Problem 1: $F(s) = \dfrac{3s + 5}{s^2 + 5s + 6}$ has one zero, $s = -\frac{5}{3}$ and two poles, $s = -2$ and $s = -3$. These may be plotted on the complex plane for s, in this case lying along the real axis (▶Fig. 6.2).

For some systems, the mathematics becomes more complicated and the roots and poles may be complex, for example:

Problem 2: $F(s) = \dfrac{s + 1}{s^2 + 2s + 2}$

The complex poles are $s = -1 \pm i$; the zero is $s = -1$. The poles lie on the left hand side of the complex plane $\mathbb{C}$. In control theory, this is a general condition for system stability, as already introduced in ▶Chap. 4.

6.7 Some Special Integrals

(i)   An integral involving splitting the denominator of the integrand into two conjugates.

$$I = \int \frac{dx}{1 - x^2}$$

Splitting the integrand into its conjugates,

$$
\begin{aligned}
\frac{1}{1 - x^2} &= \frac{1}{(1 - x)(1 + x)} \\
&= \frac{A}{1 - x} + \frac{B}{1 + x} \\
&= \frac{A(1 + x) + B(1 - x)}{1 - x^2}
\end{aligned}
$$

Comparing numerators, as usual,

$$A - B = 0$$
$$A + B = 1$$
$$\Rightarrow A = B = \frac{1}{2}$$
$$\therefore I = \frac{1}{2} \int \frac{dx}{1+x} + \frac{1}{2} \int \frac{dx}{1-x}$$
$$I = \frac{1}{2} \ln \left(\frac{1+x}{1-x} \right)$$

This method was used by Isaac Barrow to show

$$I = \int \sec \theta d\theta = \frac{1}{2} \ln \left(\frac{1 + \sin \theta}{1 - \sin \theta} \right)$$

again making use of the trigonometric form of Pythagoras' Theorem, as follows:

$$\int \sec \theta d\theta = \int \frac{\cos \theta}{\cos^2 \theta} d\theta = \int \frac{d(\sin \theta)}{\cos^2 \theta}$$
$$= \int \frac{d(\sin \theta)}{1 - \sin^2 \theta} = \int \frac{dx}{1 - x^2}$$

where $x = \sin \theta$, and which we have already solved.

(ii) $I = \int xe^{-x^2} dx$ cannot be evaluated using the linear function of a function rule, but a simple substitution works, as follows:

$$I = \int xe^{-x^2} dx = \frac{1}{2} \int e^{-x^2} d(x^2)$$
$$\therefore I = -\frac{e^{-x^2}}{2} + c$$

For the definite integral

$$I = \int_{-\infty}^{\infty} Ke^{-ax^2} dx,$$

the above method doesn't work because the factor x is absent, and a more sophisticated approach is needed. Referring to ▶Fig. 6.3, the function $f(x) = Ke^{-ax^2}$ is obtained by taking a planar section through the solid shape obtained by rotating it about the ordinate axis Oz. At this stage, attention

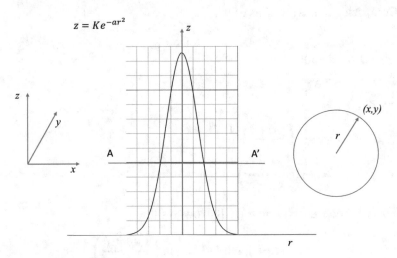

□ Fig. 6.3 Gaussian volume of revolution

should be paid to the dimensions of terms in $f(x)$.[2] For the dimensions of $f(x)$ to be length, $[f(x)] = L$, $[K] = L$ and a has units of (area)$^{-1}$ ($[a] = L^{-2}$). Then the double integral $K \int_{-\infty}^{\infty} \int_{-\infty}^{\infty} e^{-a(x^2+y^2)} dx dy$ is a true physical volume ($[V] = L^3$).

$$\therefore V = K \int_{-\infty}^{\infty} e^{-ax^2} dx \int_{-\infty}^{\infty} e^{-ay^2} dy = \frac{I^2}{K}$$

[since x and y are interchangeable "dummy" variables inside the $\int$ sign.] An alternative calculation of V uses its radial symmetry, so that sections AA' through V will be circles of radii $r^2(z) = x^2 + y^2$. V can therefore be calculated as a sum of disks:

$$V = \int_{z=0}^{K} \pi r^2(z) dz = \pi Ka \int_{0}^{\infty} -r^2 e^{-ar^2} d(r^2)$$

$$\frac{V}{\pi Ka} = -\int_{0}^{\infty} u e^{-au} du = \left[\frac{u e^{-au}}{a} \right]_{0}^{\infty} + \frac{1}{a} \int_{0}^{\infty} e^{-au} du = \frac{1}{a^2}$$

$$\Rightarrow V = \frac{\pi K}{a}$$

$$\therefore V = \frac{I^2}{K} = \frac{\pi K}{a} \Rightarrow I = K\sqrt{\frac{\pi}{a}}$$

$$\therefore \int_{-\infty}^{\infty} K e^{-ax^2} dx = K\sqrt{\frac{\pi}{a}}$$

---

2   The dimensions of variables are indicated using the square bracket notation. e.g. if m is a mass $[m] = M$. Note the distinction between dimensions (length L, time T, mass M) and units (metres, seconds, kg).

This integral occurs in Gaussian Statistics in calculations of the Error Function and Probability Intervals (see ▶Chap. 11).

6.8 Calculation of Areas

Integration is used to calculate areas under curves and enclosed by combinations of curves.

Example: Simple integration enables calculation of the area enclosed by the straight line $y = 2x$ and the curve $y = 3 - x^2$, as plotted in ▶Fig. 6.4.
The intersection between the curves sets the upper limit of integration when

$$
\begin{aligned}
3 - x^2 &= 2x \\
\therefore x^2 + 2x - 3 &= 0 \\
(x - 1)(x + 3) &= 0
\end{aligned}
$$

The intercepts are at $x = -3$ and $x = 1$ as seen in ▶Fig. 6.4.
The enclosed area A is therefore:

$$
\begin{aligned}
A &= \int_{-3}^{1} (3 - x^2)\mathrm{d}x - \int_{-3}^{1} 2x\,\mathrm{d}x \\
&= \left[3x - \frac{x^3}{3} - x^2 \right]_{-3}^{1} \\
\therefore A &= \frac{32}{3}
\end{aligned}
$$

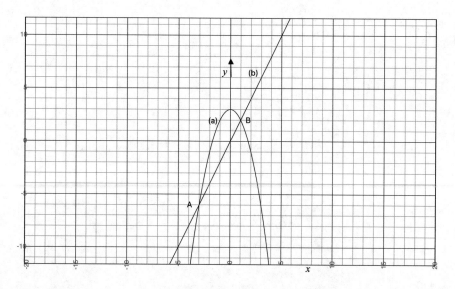

☐ **Fig. 6.4** Plots of **a** $y = 3 - x^2$ and **b** $y = 2x$ showing the area bounded by these two curves (between AB and the curve)

Functional Analysis

Contents

© The Author(s), under exclusive license to Springer Nature
Switzerland AG 2022
P. Prewett, *Foundation Mathematics for Science and Engineering Students*,
https://doi.org/10.1007/978-3-030-91963-4_7

Considering functions defined on the real line $\mathbb{R} : f(x), x \in \mathbb{R}$.

(a) Where $f(x)$ is defined $\forall x \in \mathbb{R}$

$$y = f(x), \quad -\infty < x < \infty$$

(b) Where $f(x)$ is defined on part of the Real Line

For a closed domain:

$$y = f(x), \ a \le x \le b, \qquad \text{which can also be written } x \in [a, b]$$

For an open domain:

$$y = f(x), \ a < x < b; \ x \in (a, b)$$

For an open domain (a, b) points $x = a$ and $x = b$ are not included. Note also other possibilities such as $x \in (a, b]$, and $x \in [a, b)$.

Functions defined on $\mathbb{R}$ may be monotonic or non-monotonic and may have discontinuities of different kinds (▶Fig. 7.1).

Monotonic functions can be monotonic increasing, in which for increasing $x, f(x)$ always increases. An example is the cubic function $f(x) = (x-3)^3$ (see ▶Fig. 7.2c).

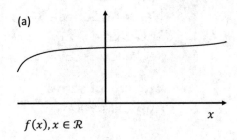

(a)

$f(x), x \in \mathcal{R}$

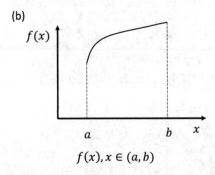

(b)

$f(x)$

$f(x), x \in (a, b)$

◻ **Fig. 7.1**   Functions defined on the Real Line (In case (b), $f(x)$ is either not defined outside (a, b) or is zero there)

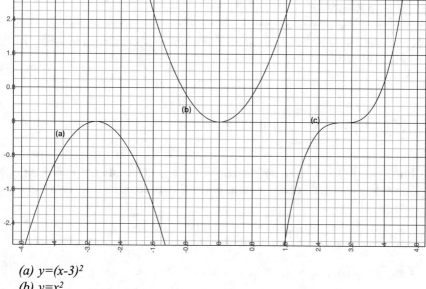

(a) y=(x-3)²
(b) y=x²
(c) y=(x-3)³

⬛ Fig. 7.2 Functions with **a** maximum, **b** minimum and **c** point of inflection

A monotonic decreasing function has $f(x)$ decreasing as x increases for example $f(x) = e^{-x}$. Non-monotonic functions are not single valued with respect to x. Examples are $f(x) = \cos x$ and $f(x) = x^2$. Non-monotonic functions display maxima and minima.

7.1 Maxima, Minima and Point of Inflection

Functions possess a peak or local maximum when $\dfrac{df}{dx} = 0$ (curve (a) in ►Fig. 7.2). But $\dfrac{df}{dx}$ is also zero at a local minimun (curve (b) ►Fig. 7.2) and at a point of inflection (curve (c)). The three cases for which $f'(x) = 0$ depends on the behaviour of the second derivative of f. Thus, for (a) $f'(x)$ decreases to zero, then becomes negative, passing through the local maximum, so that $f'' = 0$ at a maximum. From ►Fig. 7.2, the rules for identifying the three cases are:

$$
\begin{aligned}
\text{Maximum:} \quad & f'(a) = 0, \; f''(a) < 0 \\
\text{Minimum:} \quad & f'(a) = 0, \; f''(a) > 0 \\
\text{Inflection:} \quad & f'(a) = 0, \; f''(a) = 0
\end{aligned}
$$

7.2 Discontinuities

Functions may possess discontinuities. For example, function (a) in ▶Fig. 7.3 has a *point discontinuity* at $x = x_1$.

(a)

$$f(x) = \begin{cases} g(x), & x \in \mathbb{R} \text{ and } x \neq p \\ q, & x = p \end{cases}$$

(b) Function (b) has a *jump discontinuity* at $x = r$. The value of $f(r)$ is different depending on whether point r in the domain is approached from below or above along x.

$$\lim_{x \to r^-} f(x) \neq \lim_{x \to r^+} f(x)$$

(c) Function (c) has an *asymptotic discontinuity* at $x = s$

$$\lim_{x \to s^-} \to +\infty; \quad \lim_{x \to s^+} \to -\infty.$$

In this case the two branches of f only meet asymptotically at s.

(d) A continuous function with none of the above first order discontinuities may be discontinuous in its first derivative (slope) as in ▶Fig. 7.3d. This is a *second order discontinuity* for which $\lim_{x \to t^-} f'(x) \neq \lim_{x \to t^+} f'(x)$.

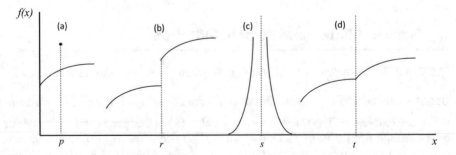

■ **Fig. 7.3** Discontinuities in functions **a** point **b** step/jump **c** asymptotic **d** second order

7.3 Tangent, Normal and Curvature

The tangent to the curve generated by a function $f(x)$ plotted on the xy plane at a point $(a, f(a))$ on the curve is the straight line $y = m_1 x + c$ where m_1 is the slope $f'(a)$ of the curve. This tangential line cuts the x−axis at angle $\theta = \tan^{-1} m_1$. The normal to the curve at $x = a$ cuts the x−axis at angle $\beta = \theta + \frac{\pi}{2}$ so that the slope of the normal at a is $m_2 = \tan(\theta + \frac{\pi}{2}) = -\cot\theta$, using ▶Fig. 1.2.
Thus, the product of the gradients of tangent and normal is $m_1 m_2 = \tan\theta(-\cot\theta) = -1$ This condition applies at any point and gives a test for functions to be mutually ⊥. Thus, for any pair of functions $f(x)$ and $g(x)$ to be mutually ⊥ (orthogonal) at $x = a$, requires

$$\frac{df}{dx}\bigg|_a \cdot \frac{dg}{dx}\bigg|_a = -1$$

Example 1: Straight lines

$$y - 3x - 5 = 0$$
$$3y + x - 6 = 0$$

$m_1 = 3, m_2 = -\frac{1}{3} \Rightarrow m_1 m_2 = -1$ proving the two lines are orthogonal.
Example 2: Find the equation of the normal to the curve
$x^2 = 4y$ at $x = 2$.

$$y = \frac{x^2}{4} \Rightarrow \frac{dy}{dx} = \frac{x}{2}$$

$$\therefore y'(2) = 1$$

The normal therefore has slope $m_2 = -1$, so that its equation is

$$y = -x + c_2$$

The normal passes through $(2, 1) \Rightarrow$

$$1 = -2 + c_2 \Rightarrow c_2 = 3$$

The equation of the normal is therefore

$$x + y = 3$$

Radius of Curvature
At any point P along a well-behaved curve, it is possible to fit a circle touching the curve which defines its local radius of curvature $R(x)$ at P as shown in ▶Fig. 7.4. This is called the Osculating Circle. The incremental distance ds along the curve in the direction of increasing x is obtained from

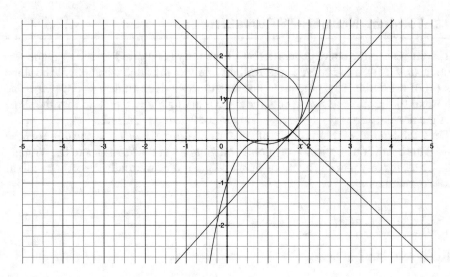

◘ Fig. 7.4 Tangent, Normal and Osculating Circle

$$\mathrm{d}s^2 = \mathrm{d}x^2 + \mathrm{d}y^2$$

$$\therefore \mathrm{d}s = \mathrm{d}x\sqrt{1 + \left(\frac{\mathrm{d}y}{\mathrm{d}x}\right)^2}$$

And, from the osculating circle,

$$\mathrm{d}s = R\mathrm{d}\theta$$

so that $R = \dfrac{\mathrm{d}x}{\mathrm{d}\theta}\sqrt{1 + \left(\dfrac{\mathrm{d}y}{\mathrm{d}x}\right)^2}$

$$\frac{\mathrm{d}y}{\mathrm{d}x} = \tan\theta \Rightarrow \frac{\mathrm{d}^2 y}{\mathrm{d}x^2} = \sec^2\theta\frac{\mathrm{d}\theta}{\mathrm{d}x}$$

$$\therefore \frac{\mathrm{d}x}{\mathrm{d}\theta} = \frac{\sec^2\theta}{y''} = \frac{1 + tan^2\theta}{y''} = \frac{1 + \left(\dfrac{\mathrm{d}y}{\mathrm{d}x}\right)^2}{y''(x)}$$

$$\therefore R(x) = \frac{1}{y''(x)}\left[1 + \left(y'(x)\right)^2\right]^{\frac{3}{2}}$$

This expression for the radius of the osculating circle is usually remembered as the Curvature $\dfrac{1}{R}$:

$$\frac{1}{R(x)} = \frac{y''}{[1+(y')^2]^{\frac{3}{2}}}$$

In many problems, as in the space curve of moving bodies, $\dfrac{dy}{dx}$ is small and the curvature is often approximated by

$$\frac{1}{R(x)} \simeq \frac{d^2 y}{dx^2}$$

to simplify calculations.

The radius of curvature is important in geometrical optics and particle motion. In 3-D dynamics the plane containing the tangent and the normal is called the Osculating Plane and the $\perp$ to it is called the Binormal.

In ►Fig. 7.4, the tangent and normal to the curve $y = (x-1)^3$ are drawn at the point $(1.6, 0.2)$. The slope of the tangent at this point is

$$\frac{dy}{dx}\bigg|_{1.6} = 3(1.6-1)^2 = 1.08$$

Its equation is $y = 1.08x + c$ and, since it passes through $(1.6, 0.2)$,

$$0.2 = 1.08 \times 1.6 + c \Rightarrow c = -1.528$$

The tangent is therefore the straight line

$$y = 1.08x - 1.528$$

Similarly, using $m_1 m_2 = -1$, the equation of the normal is

$$y = -0.925x + 1.68$$

The tangent and normal at $(1.6, 0.2)$ are drawn on the curve in ►Fig. 7.4. In order to draw the osculating circle, the radius is calculated:

$$\frac{1}{R(x)} = \frac{y''}{[1+y'^2]^{\frac{3}{2}}} = \frac{6(1.6-1)}{[1+(1.08)^2]^{\frac{3}{2}}}$$

from which $R = 0.886$.

The slope angle α of the tangent is $\tan^{-1}(1.08) = 47.2°$, from which simple geometry gives the centre of the osculating circle at distance R along the normal as

$$x_0 = 1.6 - 0.886\sin 47.2° = 0.95$$
$$y_0 = 0.2 + 0.886\cos 47.2° = 0.802$$

so that the equation of the osculating circle is

$$(x - 0.95)^2 + (y - 0.802)^2 = (0.886)^2$$

which is plotted in ▶Fig. 7.4, which shows it just grazing the curve $y = (x - 1)^3$ at $(1.6, 0.2)$.

7.4 The Binomial Theorem

A simple standard form, easy to remember, is

$$(1 + x)^n = 1 + \frac{nx}{1} + \frac{n(n-1)x^2}{1 \cdot 2} + \frac{n(n-1)(n-2)x^3}{1 \cdot 2 \cdot 3} + \cdots + x^n$$

(the dot means multiplication)

A frequently encountered alternative form can be written down from this:

$$(a + x)^n = a^n \left(1 + \frac{x}{a}\right)^n$$

The simpler form with $\frac{x}{a}$ instead of x is easier to deal with:

Example:

$$
\begin{aligned}
(x + 2)^3 &= (2 + x)^3 = 8(1 + \frac{x}{2})^3 \\
&= 8\left[1 + \frac{3x}{2} + 3\left(\frac{x}{2}\right)^2 + \left(\frac{x}{2}\right)^3\right] \\
&= x^3 + 6x^2 + 12x + 8
\end{aligned}
$$

7.4.1 Understanding the Binomial Theorem

A non-rigorous proof of the Binomial Theorem provides understanding:

$$
\begin{aligned}
1 + x &= 1 + x \\
(1 + x)^2 &= 1 + 2x + x^2 \\
(1 + x)^3 &= (1 + x)(1 + x)^2 \\
&= (1 + x)(1 + 2x + x^2) \\
&= 1 + 3x + 3x^2 + x^3 \\
(1 + x)^4 &= (1 + x)(1 + x)^3 \\
&= (1 + x)(1 + 3x + 3x^2 + x^3) \\
&= 1 + 4x + 6x^2 + 4x^3 + x^4
\end{aligned}
$$

Progressing through higher powers:

$$(1+x)^5 = (1+x)(1+x)^4$$
$$= (1+x)(1+4x+6x^2+4x^3+x^4)$$
$$= 1+5x+10x^2+10x^3+5x^4+x^5$$

$$\vdots$$

$$(1+x)^n = (1+x)(1+x)^{n-1}$$
$$= 1+nx+\frac{n(n-1)x^2}{1\cdot2}+\frac{n(n-1)(n-2)x^3}{1\cdot2\cdot3}$$
$$+\cdots+x^n \quad \text{Q.E.D.}$$

7.5 Taylor and McLaurin Series

These are important infinite series with wide ranging uses.

The curve between $x = a$ and the general point x approximated from its value $f(a)$ at $x = a$ using the straight line of slope b_1, as drawn in ▶Fig. 7.5, is given by:

$$\frac{f(x)-f(a)}{x-a} = b_1$$

Hence, the 1st order approximation to $f(x)$, knowing $f(a)$, is

$$f(x) = f(a) + b_1(x-a)$$

The 2nd order approximation is

$$f(x) = f(a) + b_1(x-a) + b_2(x-a)^2$$

and so on, including higher orders and extending this sum to infinite order gives the Taylor Series.

$$f(x) = f(a) + b_1(x-a) + b_2(x-a)^2 + b_3(x-a)^3 + \cdots$$

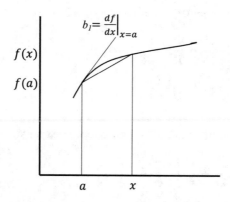

◻ **Fig. 7.5** Schematic for series approximation of a function

The coefficients b_n where $n = 1, 2, 3, \ldots$ are multiples of derivatives of f at $x = a$ of increasing order, given by

$$\left.\frac{df}{dx}\right|_{x=a} = b_1, \quad \left.\frac{d^2f}{dx^2}\right|_{x=a} = 2b_2, \quad \left.\frac{d^3f}{dx^3}\right|_{x=a} = 3 \cdot 2b_3, \ldots$$

$$\left.\frac{d^nf}{dx^n}\right|_{x=a} = n!b_n$$

The function (not approximate) is therefore given by the Taylor Series

$$f(x) = f(a) + f'(a)(x-a) + \frac{f''(a)}{2!}(x-a)^2 + \frac{f'''(a)}{3!}(x-a)^3 + \cdots$$

The special case in which the series is begun at $x = 0$ is called the McLaurin Series:

$$f(x) = f(0) + f'(0)x + f''(0)\frac{7x^2}{2!} + f'''(0)\frac{x^3}{3!} + \cdots$$

7.5.1 McLaurin Series for Key Functions

(i) $f(x) = e^x$. (Remember $\frac{d}{dx}e^x = e^x$.)

$$\therefore f(x) = 1 + x \left.\frac{de^x}{dx}\right|_0 + \frac{x^2}{2!}\left.\frac{d^2e^x}{dx^2}\right|_0 + \cdots$$

$$\Rightarrow e^x = 1 + x + \frac{x^2}{2!} + \frac{x^3}{3!} + \cdots$$

(ii) $f(x) = \sin x$
The McLaurin Series is

$$\sin x = 0 + x\cos 0 - \frac{x^2}{2!}\sin 0 - \frac{x^3}{3!}\cos 0 + \cdots$$

$$= x - \frac{x^3}{3!} + \frac{x^5}{5!} - \frac{x^7}{7!} + \cdots$$

$$\text{i.e. } \sin x = \sum_{k=0}^{\infty}(-1)^k\frac{x^{2k+1}}{(2k+1)!}$$

Exercise: Show that the 6th order McLaurin Series expansion for $\cos x$ is

$$\cos x = 1 - \frac{x^2}{2!} + \frac{x^4}{4!} - \frac{x^6}{6!} + \cdots$$

$$\text{i.e. } \cos x = \sum_{k=0}^{\infty}(-1)^k\frac{x^{2k}}{(2k)!}$$

Note: sin x contains odd orders in x with alternating sign while cos x contains only even order terms.

7.6 Example of Series Approximation

This is the base for natural logarithms. Exponential growth or decay, usually in the time domain is very important in science and engineering.

$$y = y_0 e^{at} \qquad a \text{ positive or negative}$$

e = 2.71828182... is not a rational number, i.e. it cannot be expressed as a simple fraction:

$$e \neq \frac{p}{q}$$

The Euler number e can be approximated to various orders using the McLaurin Series

$$e^x = 1 + x + \frac{x^2}{2!} + \frac{x^3}{3!} + \frac{x^4}{4!} + \cdots$$

Putting $x = 1$ to estimate e:

Order	Estimate
0	1
1	1+1=2
2	2+1/2!=2.5
3	2.5+1/3!=2.667
4	2.667+1/4!=2.7083
5	2.7083+1/5!=2.7166
6	2.7166+1/6!=2.7180
7	2.7180+1/7!=2.7182

After 7 orders, the McLaurin Series converges closely to the true value of e = 2.718281828...

7.7 Series Method for $\int \cos x^2 dx$

In physical optics this is the Fresnel Integral
The series form of cosine, for $z = x^2$, is (see ▶Sect. 7.5.1)

$$\cos z = \sum_{k=0}^{\infty} (-1)^k \frac{z^{2k}}{(2k)!}$$

$$\therefore \cos z = 1 - \frac{z^2}{2!} + \frac{z^4}{4!} - \frac{z^6}{6!} + \cdots$$

$$\int \cos x^2 dx = \int \left(1 - \frac{x^4}{2!} + \frac{x^8}{4!} - \frac{x^{12}}{6!} + \cdots\right) dx$$

$$= x - \frac{x^5}{5 \cdot 2!} + \frac{x^9}{9 \cdot 4!} - \frac{x^{13}}{13 \cdot 6!} + \cdots + c$$

$$= \sum_{k=0}^{\infty} (-1)^k \frac{x^{4k+1}}{(4k+1)(2k)!} + c$$

Exercise: Show that

$$\int \sin x^2 dx = \sum_{k=0}^{\infty} (-1)^k \frac{x^{4k+3}}{(4k+3)(2k+1)!}$$

7

7.8 The Hyperbolic Functions

These functions should not be confused with the circular functions; their names look similar but they are very different. They are formed from exponentials and, like simple exponentials, they are important in physics and engineering mathematics, in particular the mathematics of oscillations.

		Pronounced
$\sinh x$	$= \frac{1}{2}(e^x - e^{-x})$	*"shine"*
$\cosh x$	$= \frac{1}{2}(e^x + e^{-x})$	*"cosh"*
$\tanh x$	$= \frac{\sinh x}{\cosh x}$	*"than"*
$\operatorname{cosech} x$	$= \frac{1}{\sinh x}$	*"coshec"*
$\operatorname{sech} x$	$= \frac{1}{\cosh x}$	*"shec"*
$\coth x$	$= \frac{1}{\tanh x}$	*"cothan"*

Exercise:
Show that

1. $\cosh^2 x - \sinh^2 x = 1$
2. $\sinh 2x = 2 \sinh x \cosh x$

3. $\dfrac{d}{dx}(\sinh x) = \cosh x$

4. $\dfrac{d}{dx}(\cosh x) = \sinh x$

7.8.1 Hyperbolic Functions in Integration

Remember $\displaystyle\int \dfrac{dx}{\sqrt{a^2 - x^2}}$ is solved using the substitution

$x = a\sin\theta$ (►Sect. 6.5). Similarly, $\displaystyle\int \dfrac{dx}{\sqrt{x^2 + a^2}}$ is solved using the hyperbolic

function substitution $x = a\sinh z$.

$$\therefore I = \int \dfrac{\cosh z}{\sqrt{1 + \sinh^2 z}}dz$$

Using the "Hyperbolic Pythagoras Identity"
$\cosh^2 z - \sinh^2 z = 1$,

$$I = \int dz = z + c$$

$$\therefore I = \sinh^{-1}\dfrac{x}{a} + c$$

The method and result are a hyperbolic function version of the earlier trigonometric method.

Exercise: Show that

$$\int \dfrac{dx}{\sqrt{x^2 - a^2}} = \cosh^{-1}\dfrac{x}{a} + c$$

The Conic Sections

Contents

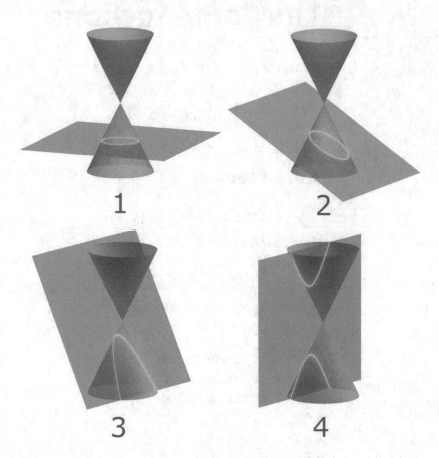

8

◨ **Fig. 8.1** Slices through cones produce the conic section 2-dimensional curves 1. Circle 2. Ellipse 3. Parabola 4. Hyperbola. ▶ https://upload.wikimedia.org/wikipedia/commons/c/cc/ TypesOfConicSections.jpg ▶ JensVyff, Own work. Creative Commons Licence CC BY-SA 4.0, via Wikimedia Commons

The conic section curves are constructed from intersections between angled planes and a 3-D solid cone to produce 2-D curves as shown in ▶ Fig. 8.1. The two "nappes" of the cone produce different branches of the conic sections. The straight line is generated from grazing incidence between cone and plane and is not universally regarded as a true conic section, but it will be included here.

8.1 Straight Line

A straight line through point (x_0, y_0) has constant slope $\forall (x, y)$, given by

$$\frac{dy}{dx} = m$$
$$\therefore y = mx + c$$

The constant of integration c is the intercept of the straight line on the y–axis ($x = 0$). c is calculated knowing any point on the line. The line passes through (x_0, y_0) from which the intercept on the y–axis is

$$c = y_0 - mx_0$$

The intercept on the x–axis ($y = 0$) is $-\dfrac{c}{m}$ (▶Fig. 8.2).

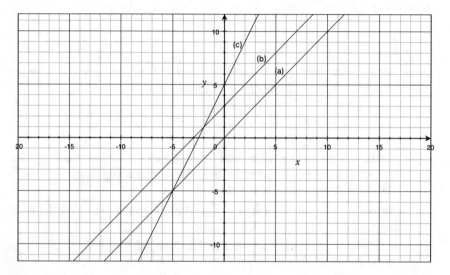

□ **Fig. 8.2** Examples of straight lines **a** $y = x$; **b** $y = x + 3$; **c** $y = 2x + 5$

8.2 Circle

The circle centered on the origin shown in ▶Fig. 8.3 is consistent with Pythagoras' Theorem:

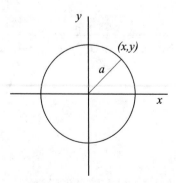

□ **Fig. 8.3** Simple circle conic section ($x^2 + y^2 = a^2$)

$$x^2 + y^2 = a^2$$

$$\left(\frac{x}{a}\right)^2 + \left(\frac{y}{a}\right)^2 = 1$$

It is obvious that a circle centred on point (x_0, y_0) instead of the origin, obeys equation

$$(x - x_0)^2 + (y - y_0)^2 = a^2$$

8.3 Ellipse

An ellipse is essentially a circle stretched along one axis

$$\left(\frac{x}{a}\right)^2 + \left(\frac{y}{b}\right)^2 = 1$$

8

If $a > b$, the ellipse's major axis is Ox, and the minor axis is Oy, as in ▶Fig. 8.4a.
If $b > a$ the ellipse major axis is in the Oy direction ((b) in ▶Fig. 8.4).
If $a = b$, the ellipse degenerates to a circle.
Referring to ▶Fig. 8.5, the directrices of the ellipse are at $x = \pm\frac{a}{e}$ on the major axis; the foci are at $x = \pm ea$ on the major axis.
Parametric definition of the ellipse in terms of its eccentricity e is:

$$\frac{PS}{PK} = e$$

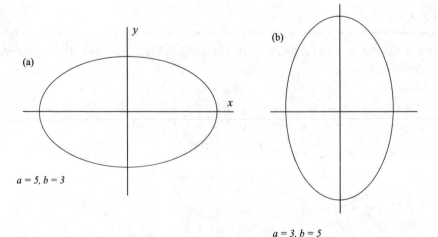

(a)

$a = 5, b = 3$

(b)

$a = 3, b = 5$

□ **Fig. 8.4** Ellipses as "stretched" circles **a** along x–axis **b** y–axis

The larger e, the greater the "stretching" of the ellipse in the Ox direction.

Analysis of the ellipse, using the methods of coordinate geometry as shown in ►Fig. 8.5, is based on the existence of foci S and S', symmetrically located on either side of the origin at points $(\pm ea, 0)$, where the eccentricity e is clearly a measure of the ellipse's deviation from circularity. The two foci are related to corresponding directrices at $x = \pm\dfrac{a}{e}$.

The ellipse is the locus of points (x, y) for which the distance from the focus is e times their distance to the corresponding directrix.

$$\frac{PS}{PK} = e \quad \Rightarrow$$

$$
\begin{aligned}
(x - ea)^2 + y^2 &= e^2(x - \frac{a}{e})^2 \\
x^2 - 2eax + e^2a^2 + y^2 &= e^2x^2 - 2eax + a^2 \\
(1 - e^2)x^2 + y^2 &= a^2(1 - e^2) \\
\frac{x^2}{a^2} + \frac{y^2}{a^2(1 - e^2)} &= 1
\end{aligned}
$$

$\dfrac{PS}{PK} = e$ must be true $\forall P(x, y)$ including at $y = \pm b$.

$\Rightarrow a^2(1 - e^2) = b^2$

So that $\dfrac{x^2}{a^2} + \dfrac{y^2}{a^2(1 - e^2)} = 1$ becomes the standard canonical equation $\dfrac{x^2}{a^2} + \dfrac{y^2}{b^2} = 1$.

$\forall$ ellipses, the eccentricity e is such that $\left(\dfrac{b}{a}\right)^2 = 1 - e^2$

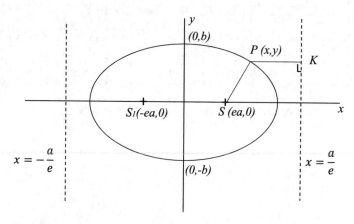

▫ Fig. 8.5 Coordinate Geometry of the Ellipse

It can be seen that a circle is an ellipse for which the eccentricity e is zero.[1]

8.4 Parabola

The standard canonical form of the parabola is $y^2 = 4ax$.
A parabola is the locus of points which are equidistant from the focus at $(a, 0)$ and the directrix at $x = -a$
i.e. $PS = PK$ $(e = 1)$

$$PS^2 = PK^2$$

i.e. $(x - a)^2 + y^2 = (x + a)^2$
$\Rightarrow \quad y^2 = 4ax$
$y = \pm 2\sqrt{ax}$.
This parabola is symmetrical about Ox with two values of y for every value of x.
A parabola symmetrical about Oy has the equation $x^2 = 4ay$ (see ▶Fig. 8.6).

8

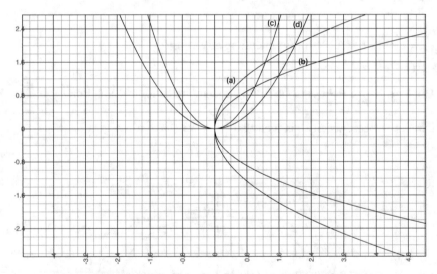

□ **Fig. 8.6** Examples of Parabolas **a** $y^2 = 2x$ **b** $y^2 = x$ **c** $x^2 = y$ **d** $x^2 = 2y$

---

1   We have used e for the eccentricity. Do not confuse this with the Euler Number, e.

8.5 Hyperbola

The canonical form is

$$\frac{x^2}{a^2} - \frac{y^2}{b^2} = 1$$

The eccentricity is $\dfrac{PS}{PK} = e > 1 \Rightarrow PS^2 = e^2 PK^2$

$$(x - ea)^2 + y^2 = e^2 \left(x - \frac{a}{e}\right)^2$$

$$\frac{x^2}{a^2} - \frac{y^2}{a^2(e^2 - 1)} = 1$$

$$\frac{x^2}{a^2} - \frac{y^2}{b^2} = 1, \quad b^2 = a^2(e^2 - 1), \quad e > 1.$$

$\exists$ 2 branches $y = \pm \dfrac{bx}{a} \sqrt{1 - \dfrac{a^2}{x^2}}$.

Note that when $x \to \pm\infty$, $y = \pm\dfrac{bx}{a}$, straight lines called the asymptotes of the hyperbola. The asymptotes are the straight lines in ▶Fig. 8.7.
The vertices are at $y = 0$

$$\frac{x^2}{a^2} = 1, \Rightarrow x = \pm a$$

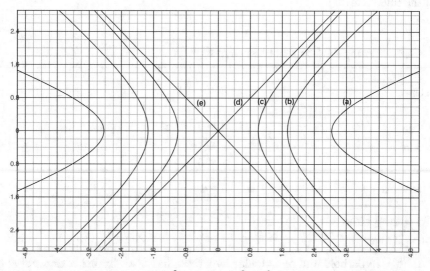

◻ **Fig. 8.7** Examples of hyperbolae **a** $\dfrac{x^2}{8} - y^2 = 1$ **b** $\dfrac{x^2}{3} - \dfrac{y^2}{2} = 1$ **c** $x^2 - y^2 = 1$ **d, e** asymptotes of (**c**)

Vertices	$x = \pm a,$	$y = 0$
Foci	$x = \pm ea,$	$y = 0$
Directrices	$x = \pm \dfrac{a}{e},$	$y = 0$

$$\text{Asymptotes} \quad y = \pm \frac{bx}{a}$$

$$\text{Eccentricity} \quad e = \sqrt{\frac{a^2 + b^2}{a^2}} > 1$$

8.6 Rectangular Hyperbola

In this special case, $b = a$ and the asymptotes $y = \pm x$ are mutually perpendicular, and may be used as axes $x'y'$ tilted at 45° to the xy axes. This is the case in ▶Fig. 8.7 for curve (c) $x^2 - y^2 = 1$ which is a rectangular hyperbola.

The general point $P(x, y)$ on the hyperbola becomes $P(x', y')$ in the new coordinate system, such that

$$x = y' \cos \tfrac{\pi}{4} + x' \cos \tfrac{\pi}{4} = \tfrac{1}{\sqrt{2}}(y' + x')$$
$$y = y' \sin \tfrac{\pi}{4} - x' \sin \tfrac{\pi}{4} = \tfrac{1}{\sqrt{2}}(y' - x')$$

$$\therefore x'y' = \frac{1}{2}(x^2 - y^2) = \frac{a^2}{2}$$

which is the simple form of the equation for the rectangular hyperbola in the frame of the tilted axes $x'y'$.

8.7 Summary of Canonical Forms

Ellipse	$\dfrac{x^2}{y^2} + \dfrac{y^2}{b^2} = 1$	$0 < e < 1$
Circle	$\dfrac{x^2}{a^2} + \dfrac{y^2}{a^2} = 1$	$e = 0$
Parabola	$y^2 = 4ax$	$e = 1$
	(2 branches)	$x < 0, x > 0$
Hyperbola	$\dfrac{x^2}{a^2} - \dfrac{y^2}{b^2} = 1$	$e > 1$
Rectangular Hyperbola	$xy = \dfrac{a^2}{2}$	
Straight Line	$y = mx + c$	

Parabola, hyperbola and rectangular hyperbola have two branches depending on whether x is positive or negative.

8.8 Perimeter, Area and Volume

8.8.1 Right Angled Triangle

Straight lines can form areas, for example a triangle formed from 3 non-parallel lines.

(a)
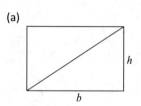

(b)
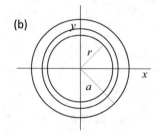

(c)
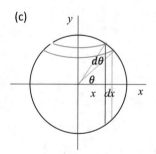

(d)
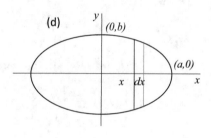

■ **Fig. 8.8** Calculation of areas and volumes **a** triangle **b** circle **c** sphere **d** ellipse

From ▶Fig. 8.8a,

$$\text{Area of right triangle} = \tfrac{1}{2} \text{ area of rectangle}$$

$$= \tfrac{1}{2} \text{ base} \times \text{height}$$

$$= \tfrac{1}{2}bh$$

The perimeter is $b + h + \sqrt{b^2 + h^2}$

8.8.2 Circle

From ▶Fig. 8.8b, the arc length is

$$
\begin{aligned}
\mathrm{d}L &= a\mathrm{d}\theta \quad \text{for } \theta \text{ measured in radians} \\
L &= a \int_0^{2\pi} \mathrm{d}\theta \\
\text{Perimeter L} &= 2\pi a
\end{aligned}
$$

$$\mathrm{d}A = 2\pi r \mathrm{d}r$$

$$A = 2\pi \int_0^a r\mathrm{d}r = \pi a^2$$

8.8.3 Sphere

From ►Fig. 8.8c

$$
\begin{aligned}
\mathrm{d}A &= 2\pi a \cos\theta \, a \, \mathrm{d}\theta \\
A &= 2\pi a^2 \int_{-\frac{\pi}{2}}^{\frac{\pi}{2}} \cos\theta \, \mathrm{d}\theta = 4\pi a^2 \\
\mathrm{d}V &= \pi y^2 \mathrm{d}x \\
V &= \pi \int_{-a}^{a} y^2 \mathrm{d}x = \pi \int_{-a}^{a} (a^2 - x^2) \mathrm{d}x \\
&= 2\pi \int_0^a (a^2 - x^2) \mathrm{d}x \quad \text{(by symmetry)} \\
&= 2\pi \left[a^2 x - \frac{x^3}{3} \right]_0^a = \frac{4}{3}\pi a^3
\end{aligned}
$$

8.8.4 Ellipse

From ►Fig. 8.8d
Area:

$$
\begin{aligned}
\mathrm{d}A &= 2y\mathrm{d}x \\
A &= 4b \int_0^a \sqrt{1 - \frac{x^2}{a^2}} \mathrm{d}x
\end{aligned}
$$

$$\text{Substituting } x = a\sin\theta \quad \mathrm{d}x = a\cos\theta \mathrm{d}\theta$$

$$
\begin{aligned}
A &= 4ba \int_0^{\frac{\pi}{2}} \cos^2\theta \mathrm{d}\theta \\
&= 2ab \int_0^{\frac{\pi}{2}} (1 + \cos 2\theta)\mathrm{d}\theta \\
&= 2ab \left[\theta + \frac{1}{2}\sin 2\theta \right]_{\theta=0}^{\frac{\pi}{2}}
\end{aligned}
$$

$$A = \pi ab$$

Check: when $a = b$ the ellipse is a circle and $A = \pi a^2$ as required.

Volume: $V = \displaystyle\int_{-a}^{a} \pi y^2 dx$ is the Volume of Revolution of the ellipsoid generated by rotating the ellipse about its major axis---the x-axis in ▶Fig. 8.4.

Through symmetry

$$V = 2\pi b^2 \int_0^a \left(1 - \frac{x^2}{a^2}\right) dx$$

$$= 2\pi b^2 \left[x - \frac{x^3}{3a^2}\right]_0^a$$

$$= \frac{4\pi}{3} ab^2$$

When $a = b$ (sphere), $V = \frac{4}{3}\pi a^3$.

Exercise:

1. Show that the area of a cone of base radius a and height h is

$$\pi a^2 \left[1 + \frac{\sqrt{a^2 + h^2}}{a}\right]$$

2. Show that the volume of revolution of the cone is

$$V = \frac{\pi h^3 d^2}{12\left[(\frac{d}{2})^2 + h^2\right]}$$

3. Show that the volume of revolution of a parabola of canonical form on the domain $x \in (0, b)$ is $2\pi ab^2$

It is notoriously difficult to produce a formula for the perimeter of an ellipse and the problem has exercised some of the best mathematical minds. Ramanujan's Formula for the perimeter is

$$L = \pi(a + b)\left(1 + \frac{3h}{10 + \sqrt{4 - h}}\right)$$

where

$$h = \frac{(a - b)^2}{(a + b)^2}.$$

1st Order Differential Equations

Contents

© The Author(s), under exclusive license to Springer Nature
Switzerland AG 2022
P. Prewett, *Foundation Mathematics for Science and Engineering Students*,
https://doi.org/10.1007/978-3-030-91963-4_9

A 1st order ODE[1] with constant coefficients has the standard form

$$\frac{dy}{dx} + ay = f(x)$$

Example:

$$\frac{dy}{dx} + 3y = x$$

which is solved by multiplying by an Integrating Factor e^{3x}, so that

$$e^{3x}\frac{dy}{dx} + 3e^{3x}y = xe^{3x}$$

$$\therefore \frac{d}{dx}(e^{3x}y) = xe^{3x}$$

The equation is now in integrable form

$$\int d(e^{3x}y) = \int xe^{3x}dx$$

9

Integrating by parts,

$$\int xe^{3x}dx = [\frac{xe^{3x}}{3}]_0^x - \frac{1}{3}\int e^{3x}dx$$

$$\therefore e^{3x}y = \frac{x}{3}e^{3x} - \frac{e^{3x}}{9} + c$$

$$y(x) = \frac{x}{3} + ce^{-3x} - \frac{1}{9}$$

The solution obtained is only defined within an unknown arbitrary constant. Unless the problem provides us with a so-called constraint to enable us to fix a value for C, the solution is an infinite family of functions or curves, corresponding to the infinite range of values C can take. This uncertainty disappears, for example, if we are told that $y = 0$ when $x = 0$. This constraint places a boundary condition on the problem, which may be referred to as the initial condition if the problem is in the time domain.

Applying this to the solution so far,

$$C = \frac{1}{9}$$

The full solution is, therefore

$$y(x) = \frac{x}{3} + \frac{e^{-3x}}{9} - \frac{1}{9}$$

---

1   A 1st order Partial Differential Equation (PDE) is linear in the first partial derivatives $\frac{\partial y}{\partial t}$, $\frac{\partial y}{\partial x}$ if $y = y(x, t)$.

As a check, this is differentiated and substituted in the ODE

$$\frac{1}{3} - \frac{3e^{-3x}}{9} + \frac{3x}{3} + \frac{3e^{-3x}}{9} - \frac{3}{9} = x$$

which reduces to $x = x$ as required.

Example:

$$\frac{dy}{dt} - 2y = 2e^{3t}; \quad y = 2 \text{ when } t = 0$$

Note that the right hand side is a function of t.

Multiply by the Integrating Factor $e^{-\int 2dt} = e^{-2t}$

$$e^{-2t}\frac{dy}{dt} - 2e^{-2t}y = 2e^{t}$$

$$e^{-2t}y(t) = 2\int e^{t}dt$$

$$e^{-2t}y(t) = 2e^{t} + c$$

Using the initial condition,

$$2 = 2 + c \Rightarrow c = 0$$

$$y(t) = 2e^{3t}$$

9.1 Summary of General Methods

The general form of a 1st order ODE with constant coefficients is

$$\frac{dy}{dt} + py = f(t)$$

When $f(t) \neq 0$, solve using the Integrating Factor $e^{\int pdt}$.
An initial condition $y(0)$ or a more general condition $y(t_0)$ is needed to evaluate the constant of integration.
In the case of the Homogeneous Equation when $f(t) = 0$, the method of separation of the variables may be used.

$$\frac{dy}{dt} + py = 0; \quad y = y_0 \text{ when } t = 0$$

$$\int \frac{dy}{y} = -\int pdt + c$$

$$\ln y = -pt + c$$

Initial condition $\Rightarrow c = \ln y_0$ Therefore $y = y_0 e^{-pt}$.

This is the classic form of exponential decay, for example when a radioactive nucleus emits alpha particles.

The Integrating Factor method is simple and always works. But another method gives an insight into the theory of differential equations and leads into the general method for solving 2nd order ODEs (with a term in $\dfrac{d^2y}{dt^2}$ or $\dfrac{d^2y}{dx^2}$).

General method for the solution of a 1st order Ordinary Differential Equation of inhomogeneous form (rhs $\neq 0$):

$$\frac{dy}{dx} + p(x)y = f(x)$$

The complementary function $y_{CF}(x)$ is the solution of the homogeneous equation

$$\frac{dy}{dx} + p(x)y = 0$$

The particular integral y_{PI} is any function which is *a* solution of the inhomogeneous equation.

The general solution is

$$y(x) = y_{CF}(x) + y_{PI}(x)$$

Proof:

$$\left[\frac{d}{dx} + p(x)\right] y(x) = f(x)$$

$$\left[\frac{d}{dx} + p(x)\right](y_{CF}(x) + y_{PI}(x)) = 0 + \left[\frac{d}{dx} + p(x)\right] y_{PI}(x)$$

$$= f(x).$$

Example:

$$\frac{dy}{dx} + 4y = 2; \text{ given } y(0) = 0$$

Homogeneous Equation

$$\frac{dy}{dx} + 4y = 0 \Rightarrow \int \frac{dy}{y} = -\int 4dx$$

$$\therefore y_{CF} = ce^{-4x}$$

Trial PI: $\qquad\qquad\qquad y_{PI} = a$

$$0 + 4a = 2 \Rightarrow a = \frac{1}{2}$$

$$\therefore y_{PI} = \frac{1}{2}$$

Complete Solution

$$y(x) = y_{CF} + y_{PI}$$

$$y(x) = ce^{-4x} + \frac{1}{2}$$

Applying the boundary condition,

$$0 = c + \frac{1}{2} \Rightarrow c = -\frac{1}{2}$$

$$y(x) = \frac{1}{2}(1 - e^{-4x})$$

Choosing a trial solution to obtain a particular integral is relatively straightforward, using the following table (▶Table 9.1).

Example:

$$\frac{dy}{dx} + 9y = 3x^2; \quad y(0) = 0$$

$$
\begin{aligned}
\text{CF} \quad \frac{dy_{CF}}{dx} + 9y_{CF} &= 0 \\
\int \frac{dy_{CF}}{y_{CF}} &= -9 \int dx + \ln A \\
y_{CF}(x) &= Ae^{-9x}
\end{aligned}
$$

☐ **Table 9.1** Selected trial particular integrals

$f(x)$	PI to try
k	a
kx	$ax + b$
kx^2	$ax^2 + bx + c$
ke^{px}	ae^{px}
$k \cos \omega x$	$a \cos \omega x + b \sin \omega x$
$k \cos \omega x + p \sin \omega x$	$a \cos \omega x + b \sin \omega x$

Note: we cannot apply the boundary condition to y_{CF} as it only fits the full solution. So, at this stage, A is unknown.

From the table, try a **PI** of form $y_{PI}(x) = ax^2 + bx + c$. Inserting into the inhomogeneous equation,

$$2ax + b + 9(ax^2 + bx + c) = 3x^2$$
$$9ax^2 + (9b + 2a)x + b + 9c = 3x^2$$

from which

$$a = \frac{1}{3}$$

$$9b + \frac{2}{3} = 0 \Rightarrow b = -\frac{2}{27}$$

$$-\frac{2}{27} + 9c = 0 \Rightarrow c = \frac{2}{9 \cdot 27}$$

$$y_{PI}(x) = \frac{x^2}{3} - \frac{2}{27}x + \frac{2}{9 \cdot 27}$$

and the general solution is

$$y(x) = y_{CF}(x) + y_{PI}(x)$$

$$y(x) = Ae^{-9x} + \frac{x^2}{3} - \frac{2}{27}x + \frac{2}{9 \cdot 27}$$

Applying the boundary condition $y(0) = 0$,

$$A = -\frac{2}{9 \cdot 27}$$

$$y(x) = \frac{x^2}{3} - \frac{2}{27}x + \frac{2}{9 \cdot 27}(1 - e^{-9x})$$

This is not a simple method compared with the alternative integrating factor approach, presented below as a comparison, but it represents a powerful rigorous general method which is essential for solving second order ODEs of form $\frac{d^2y}{dx^2} + a\frac{dy}{dx} + by = f(x)$ (see ▶Chap. 10). The same equation solved using the integrating factor method gives:

$$\frac{dy}{dx} + 9y = 3x^2$$

$$\int d(e^{9x}y) = \int 3x^2 e^{9x}dx + c$$

$$= \frac{1}{3}\left[x^2 e^{9x}\right]_0^x - \frac{2}{3}\int xe^{9x}dx$$

$$= \frac{1}{3}\left[x^2e^{9x}\right]_0^x - \frac{2}{3}\left\{\left[\frac{xe^{9x}}{9}\right]_0^x - \frac{1}{9}\int_0^x e^{9x}dx\right\}$$

$$e^{9x}y(x) = \frac{1}{3}x^2e^{9x} - \frac{2}{27}xe^{9x} + \frac{2}{9\cdot 27}e^{9x} + c$$

$$y(x) = \frac{1}{3}x^2 - \frac{2}{27}x + \frac{2}{9\cdot 27} + ce^{-9x}$$

$$y(0) = 0 \Rightarrow c = -\frac{2}{9\cdot 27}$$

$$y(x) = \frac{x^2}{3} - \frac{2}{27}x + \frac{2}{9\cdot 27}(1 - e^{-9x}) \text{ as before}$$

This method is simple in principle but, in this case, requires integration by parts to be applied twice sequentially.

9.2 Capacitor Discharge

A capacitor is a passive electrical component which is used to store electrical charge. In the circuit below, the capacitor C is charged up with a charge Q_0. When the switch, S, is closed, the capacitor is discharged through the resistor R. The current through the resistor falls to zero with time and is measured by the ammeter A; the voltage drop across the resistor at any instnt is measured by the voltmeter V (▶Fig. 9.1). Solving this problem requires physics to be used to obtain mathematical equations which are then solved to describe what happens when S is closed at time zero. Definition of capacitance C:

$$C = \frac{Q}{V} \tag{9.1}$$

where C is in Farads, Q Coulombs, V volts. There are two unknowns in this equation, viz. $Q(t)$, $V(t)$; C is a constant. A second equation is therefore required. Ohm's Law gives

$$V(t) = i(t)R \tag{9.2}$$

where i (Amps) is the instantaneous current.

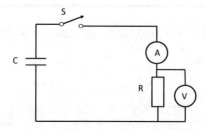

◻ Fig. 9.1 Discharging a capacitor

Since $i(t) = -\dfrac{\mathrm{d}Q}{\mathrm{d}t}$, the rate of flow of charge, from ▶(9.1),

$$i = -C\frac{\mathrm{d}V}{\mathrm{d}t} \tag{9.3}$$

▶(9.2) and ▶(9.3) $\Rightarrow$

$$\frac{\mathrm{d}V}{\mathrm{d}t} + \frac{V}{RC} = 0 \tag{9.4}$$

Solving by separation of variables (since rhs = 0).

$$\int_{V_0}^{V(t)} \frac{\mathrm{d}V}{V} = -\frac{1}{RC}\int_0^t \mathrm{d}t$$

$$\Rightarrow \ln\frac{V}{V_0} = -\frac{t}{RC}$$

$$\Rightarrow V = V_0 \mathrm{e}^{-\frac{t}{RC}}$$

The voltage on the capacitor falls with time after S is closed, with an e-folding time of RC. This is called the time constant of the circuit.
e.g. For $C = 1\mathrm{nF}$, $R = 10\,\mathrm{m}\Omega$, $RC = 10\,\mathrm{ms}$

9

9.3 Understanding Exponential Growth

Many "real-world" problems involve exponential growth with time. For example, a problem of considerable importance at the time of writing this book is growth of Covid-19 infection in the population. The exponential function is

$$y = \mathrm{e}^x \text{ where e} = 2.718281\ldots$$

Consider a viral pandemic in which every person of the infected population $N(t)$, measured at time t since the pandemic began, will infect others at the rate of a per day. The increase in the number infected in a time interval $t \to t + \mathrm{d}t$ days will be

$$\mathrm{d}N = Na\mathrm{d}t$$

$$\int \frac{\mathrm{d}N}{N} = a\int \mathrm{d}t$$

Hence, $\ln N = at + c$.
If we start counting from time $t = 0$ when the infected population was N_0,

$$\ln N = at + \ln N_0$$

$$\ln\left(\frac{N}{N_0}\right) = at$$

$$N = N_0 e^{at}$$

The number infected grows exponentially with time.

In real situations the time for which a person is infected is finite (typically a few weeks) and, sadly, some people will die, both of which mitigate against pure exponential growth in the pandemic.

2nd Order Differential Equations

Contents

P. Prewett, *Foundation Mathematics for Science and Engineering Students*,
https://doi.org/10.1007/978-3-030-91963-4_10

10.1 General Solution

We consider only ODEs with constant coefficients

$$\frac{d^2y}{dx^2} + a\frac{dy}{dx} + by = f(x)$$

Two boundary conditions are required to determine the 2 constants of integration when this equation is solved[1]

  e.g. $y = 0$, $y' = 0$ when $x = 0$

The homogeneous form is when $f(x) = 0$

$$\frac{d^2y}{dx^2} + a\frac{dy}{dx} + by = 0$$

In general, this can have a number of independent solutions $y_1(x)$, $y_2(x)$, ..., $y_n(x)$. When any number of independent solutions are added together in linear combination, the result is also a solution, since

$$c_1\frac{d^2y_1}{dx^2} + c_1a\frac{dy_1}{dx} + c_1by_1 = 0$$

$$c_2\frac{d^2y_2}{dx^2} + c_2a\frac{dy_2}{dx} + c_2by_2 = 0$$

$$\Rightarrow \frac{d^2}{dx^2}(c_1y_1 + c_2y_2) + a\frac{d}{dx}(c_1y_1 + c_2y_2) + b(c_1y_1 + c_2y_2) = 0$$

and so on for any linear combination of all possible results. The solution of a homogeneous 2nd order ODE can be understood by an example:
Example:

$$\frac{d^2y}{dx^2} + 3\frac{dy}{dx} + 2y = 0,$$

  given boundary conditions $y = 1$, $y' = -1$, when $x = 0$

Taking a trial solution: $y = e^{px}$,

$$p^2y + 3py + 2y = 0$$
$$\Rightarrow p^2 + 3p + 2 = 0$$

This is called the auxiliary equation. Its roots are $p_1 = -1, p_2 = -2$. Using the linear combination principle, the general solution is a linear combination of e^{-x} and e^{-2x}

---

1   One boundary or initial condition is sufficient when solving 1st order ODEs, as seen in ►Chap. 9.

$$y(x) = Ae^{-x} + Be^{-2x}$$

A, B are determined from the initial conditions

$$y(0) = 1 \Rightarrow A + B = 1$$

$$y'(0) = 0 \Rightarrow A + 2B = 1$$

$\therefore$ The final solution is $A = 1, \ B = 0$

$$y(x) = e^{-x}$$

10.2 Case of Repeated Roots

$$\frac{d^2y}{dx^2} - 4\frac{dy}{dx} + 4y = 0 \quad y(0) = 0; \ y'(0) = 2$$

Auxiliary equation:

$$p^2 - 4p + 4 = 0$$
$$p_1 = p_2 = 2$$

In this case, the most general solution is of the form $y(x) = (A + Bx)e^{2x}$

$$\left. \begin{array}{rcl} y(0) &=& A = 0 \\ y'(0) &=& 2 = A + B \end{array} \right\} B = 2$$

$$y(x) = 2xe^{2x}$$

10.3 Case of Complex Roots

$$\frac{d^2y}{dx^2} + 2\frac{dy}{dx} + 5y = 0 \quad y(0) = 1, y'(0) = 0$$

Auxiliary equation:

$$p^2 + 2p + 5 = 0$$

$$p_1, p_2 = -1 \pm 2i$$

General solution:

$$y(x) = Ae^{p_1 x} + Be^{p_2 x} = e^{-x}[Ae^{i2x} + Be^{-i2x}]$$

Euler's formula $\Rightarrow y(x) = e^{-x}[(A + B)\cos 2x + i(A - B)\sin 2x]$. Simplifying,

$$y(x) = e^{-x}[C\cos 2x + D\sin 2x]$$

where C, D may be complex.
Applying the boundary conditions

$$y(0) = (C\cos 0 + D\sin 0) = 1$$
$$C = 1$$

$$y'(0) = 0 \Rightarrow C - 2D = 0 \Rightarrow D = \frac{1}{2}$$

$$y(x) = e^{-x}\left[\cos 2x + \frac{1}{2}\sin 2x\right]$$

Exercise: Check by substitution in the ODE.

10.4 2nd Order ODEs in Applications

Second order ODEs are essential for solving problems involving oscillating mechanical systems and electrical oscillators in classical engineering science. An example is the spring-mass-damper system on a horizontal plane shown in ▶Fig. 10.1.

10

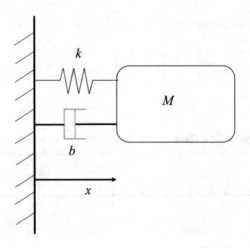

◘ **Fig. 10.1** Schematic of spring mounted sliding mass with hydraulic damper

The mass deflected through distance x is subjected to a restoring force due to the spring extension

$$F_S = -kx$$

During motion, there is a velocity dependent dissipative (non-conservative) damping force

$$F_D = -bx$$

By Newton's 2nd Law of Motion

$$M\ddot{x} = F_S + F_D = -kx - b\dot{x}$$

$$\ddot{x} + \frac{b}{M}\dot{x} + \frac{k}{M}x = 0$$

At $t = 0$, $x = x_0$, $\dot{x} = 0$ (starting from rest)

The auxiliary equation $\left(\text{with } \alpha = \dfrac{b}{M} \text{ and } \omega = \sqrt{\dfrac{k}{M}}\right)$ is

$$p^2 + \alpha p + \omega^2 = 0$$

$$\text{So that } p_1, p_2 = \frac{-\alpha \pm \sqrt{\alpha^2 - 4\omega^2}}{2}$$

Case 1, no damping:

$$\alpha = 0, \quad p_1, p_2 = \pm i\omega$$

$$x(t) = A\cos\omega t + B\sin\omega t$$

$$x = x_0, t = 0 \Rightarrow x(t) = x_0\cos\omega t + B\sin\omega t$$

$$\dot{x}(0) = 0 \Rightarrow B = 0$$

$$\therefore x(t) = x_0\cos\omega t$$

This is simple harmonic motion with amplitude x_0 and frequency $\omega = \sqrt{\dfrac{k}{M}}$.

Case 2, light damping:

$$\alpha^2 < 4\omega^2$$

$$p_1, p_2 = -\frac{\alpha}{2} \pm i\Omega$$

where $\Omega = \sqrt{\omega^2 - \dfrac{\alpha^2}{4}}$

$$x(t) = e^{-\frac{\alpha}{2}t}(A\cos\Omega t + B\sin\Omega t)$$

Initial conditions $\Rightarrow$

$$x(t) = x_0 e^{-\frac{\alpha}{2}t}\left(\cos\Omega t + \frac{\alpha}{2\Omega}\sin\Omega t\right)$$

The exponential damping of the oscillation derives from the velocity dependent dashpot damping. The frequency Ω of the oscillation is shifted from the undamped simple harmonic value ω by a damping dependent term to give a reduced oscillation frequency $\Omega = \dfrac{1}{2}\sqrt{4\omega^2 - \alpha^2}$. The form of the damped oscillation is as indicated in ►Fig. 10.2.

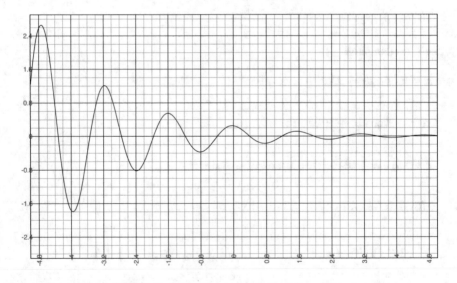

■ **Fig. 10.2**   Oscillation of lightly damped spring---mass system (time variation)

Case 3, critical damping: When $\alpha = 2\omega$ the auxiliary equation has repeated real roots

$$p_1 = p_2 = -\frac{\alpha}{2}$$

and $x(t) = (A + Bt)e^{-\frac{\alpha}{2}t}$.

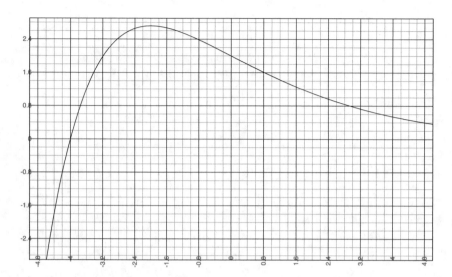

Fig. 10.3 Plot of critical damping for typical spring-mass system (time variation)

Initial conditions $\Rightarrow x(t) = x_0 \left(1 + \frac{\alpha}{2}t\right) e^{-\frac{\alpha t}{2}}$ The solution is non-oscillating as indicated in ▶Fig. 10.3. The stronger the damping α, the more rapid the decay of the transient response. This is the situation preferred for moving coil galvanometer and other scientific instruments where oscillation is undesirable. The above describes "unforced" 2nd order ODEs with constant coefficients, which are homogeneous. Inhomogeneous 2nd order ODEs, where the right hand side is non-zero, arise in physics when oscillations are driven by external forces which may or may not be oscillating. The governing equation is then of the form

$$\ddot{x} + \alpha\dot{x} + \omega^2 x = \frac{F(t)}{M}$$

A simple example would be the spring mass system in ▶Fig. 10.1 tilted vertically, whereupon the governing equation includes the acceleration due to gravity g.

$$\ddot{x} + \alpha\dot{x} + \omega^2 x = g.$$

Solutions of inhomogeneous cases are beyond the scope of this book.

Gaussian Statistics

Contents

© The Author(s), under exclusive license to Springer Nature
Switzerland AG 2022
P. Prewett, *Foundation Mathematics for Science and Engineering Students*,
https://doi.org/10.1007/978-3-030-91963-4_11

11.1 Basic Statistics

The bar graph ▶Fig. 11.1 plots the frequency $f(x)$ of results for the measurand x lying in the range $x \to x + \Delta x$. This is called the Statistical Distribution of x. Consider a set of 10 measurements yielding the following results for x

x	Frequency f
2-4	1
4-6	2
6-8	3
8-10	2
10-12	2

Taking the mid-point of each interval since the measurement does not give greater precision, the most probable result, or *mode*, of the distribution is $x_0 = 7$.

The *mean* is obtained by multiplying each value by the number of times it occurs, adding up all the results and dividing by the total number of measurements. This is the same as multiplying each result by the probability of it occurring and summing.

$$\bar{x} = \frac{1 \times 3 + 2 \times 5 + 3 \times 7 + 2 \times 9 + 2 \times 11}{3 + 2 + 2 + 2 + 1} = \frac{74}{10} = 7.4$$

This can be generalized for any distribution of n intervals with mid-interval values x_r of frequency f_r

11

$$\bar{x} = \frac{\sum\limits_{r=1}^{n} f_r x_r}{\sum\limits_{r=1}^{n} f_r}$$

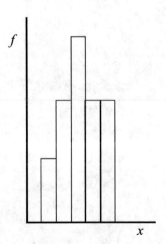

☐ **Fig. 11.1**   Bar chart showing frequencies of results

$$\sum_{r=1}^{n} f_r = N \text{ is the total number of measurements made.}$$

Three more important statistical quantities are:

(i) Root mean square value of the measurand

$$x_{rms} = \sqrt{\left[\frac{1}{N} \sum_{r=1}^{n} x_r^2 f_r \right]} = 7.81$$

(ii) Variance

$$v = \frac{1}{N} \sum_{r=1}^{n} (x_r - \bar{x})^2 f_r = 6.24$$

(iii) Standard Deviation

$$\sigma = \sqrt{v} = 2.498$$

11.2 Continuous Distribution

The Gaussian or Normal Distribution is the most important continuous statistical distribution of measured results. Think of it as a bar graph for which the measurement interval $\Delta x \to 0$.

Often called the Bell Curve, the Gaussian Distribution obeys the equation

$$f_G(x) = \frac{1}{\sigma\sqrt{2\pi}} e^{\frac{-(x-x_0)^2}{2\sigma^2}}$$

It's a probability density function: the probability of results in the range $x \to x+dx$ is $f_G(x) = dx$

x_0 = most probable value or mode.

σ = standard deviation.

$f_G(x)$ is plotted in ▶Fig. 11.2 for three different combinations of mode and variance: The Standard Normal Form, plotted at (d) is discussed below in ▶Sect. 11.3

(a) $x_0 = 1, \sigma^2 = 1$
(b) $x_0 = 1, \sigma^2 = 0.1$
(c) $x_0 = 0, \sigma^2 = 0.1$

(d) $g(z) = \frac{1}{\sqrt{2\pi}} e^{\frac{1}{2}z^2}$ (Standard Normal)

The smaller σ, the less the spread of results and the narrower the curve.

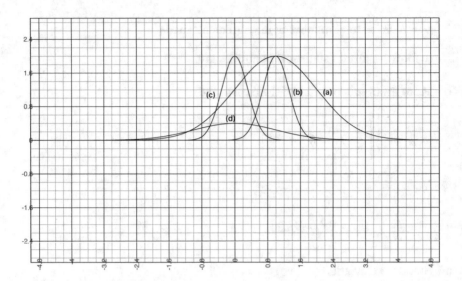

◘ Fig. 11.2 Schematic Gaussian curves

Note the units of f_G are $\left[\dfrac{1}{\sigma}\right]$ which are $[x]^{-1}$. If x is a length $[f_G]$ is metres^{-1}; if x is a statistical time delay $[f_G]$ is s^{-1}.

11.3 Standard Normal Form

This form of the Gaussian Distribution is easier to work with, using the modified measurand $z = \dfrac{x - x_0}{\sigma}$ and produces the Standard Normal Distribution from the Gaussian

$$g(z) = \frac{1}{\sqrt{2\pi}} e^{-\frac{1}{2}z^2}.$$

The probability of a result in the range $\to z + dz$ is

$$p(z) = \frac{1}{\sqrt{2\pi}} e^{-\frac{1}{2}z^2} dz$$

or $p(z) = g(z)dz$

The Bell Curve shape is maintained, but $g(z)$ is symmetrical about $z = 0$ (curve (d) in ▶Fig. 11.2).

The mode of the distribution is $z = 0$ ($x = x_0$) and the mean (by symmetry) is also $z = 0$.

The probability of a result lying in the range $-\infty < z < \infty$ must be 100% which is verified by integrating $g(x)$:

$$\frac{1}{\sqrt{2\pi}} \int_{-\infty}^{\infty} e^{-\frac{1}{2}z^2} dz = 1$$

(We showed how to calculate this integral in ▶Sect. 6.7)
$e^{-\frac{1}{2}}$ fall off of $g(z)$ on either side of the peak at $z = 0$, occurs when

$$z = \pm 1$$

Hence, the standard deviation of the Standard Normal Distribution $g(z)$ is always 1.

$f_G(z)$ varies in width depending on σ whereas the Standard Normal Form $g(z)$ has fixed width.

A lesser used width for the Gaussian curve is the full width at half maximum (FWHM). The two half maxima occur when

$$g(z) = \frac{1}{2}g_{max} = \frac{1}{2} \cdot \frac{1}{\sqrt{2\pi}}e^{-\frac{1}{2}z^2}\bigg|_{z=0}$$

$$\frac{1}{\sqrt{2\pi}}e^{-\frac{1}{2}z^2} = \frac{1}{2\sqrt{2\pi}}$$

$$e^{\frac{1}{2}z^2} = 2 \Rightarrow z^2 = 2\ln 2$$

$$z = \pm\sqrt{2\ln 2} \Rightarrow \text{FWHM} = 2\sqrt{2\ln 2}$$

In the x-domain, FWHM $= 2\sigma\sqrt{2\ln 2}$.

11.4 Confidence Intervals

Probability of a result in the range $-R < z < R$ is

$$P_{\pm R} = \int_{-R}^{R} g(z)\mathrm{d}z = 2\int_0^R g(z)\mathrm{d}z$$

$$\therefore P_{\pm R} = \sqrt{\frac{2}{\pi}}\int_0^R e^{-\frac{1}{2}z^2}\mathrm{d}z$$

Using tabulated results:

z-domain	x-domain	$P_{\pm R}$
± 1	$\pm\sigma$	0.683
± 2	$\pm 2\sigma$	0.954
± 3	$\pm 3\sigma$	0.997

68.3% of all results lie within $\pm\sigma$ of the mean, rising to 99.7% of all results for a range $\pm 3\sigma$ on either side of the mean.

11.5   The Error Function erf z

$$P_{\pm z} = \sqrt{\frac{2}{\pi}} \int_0^z e^{-\frac{1}{2}z^2} dz$$

Let $t = \dfrac{z}{\sqrt{2}}$, then $P_{\pm z} = \dfrac{2}{\sqrt{\pi}} \int_0^z e^{-t^2} dt$. $P_{\pm z} = \mathrm{erf}\, z$ where $\mathrm{erf}\, z = \dfrac{2}{\sqrt{\pi}} \int_0^z e^{-t^2} dt$ is the probability that a result lies within the range $\pm z$.

The Complementary Error Function $\mathrm{erf}_c z$ gives the probability that a result lies *outside* the range $\pm z$, and, since the two probabilities must add up to 1,

$$\mathrm{erf}_c z = 1 - \mathrm{erf}\, z.$$

11

Glossary

© The Author(s), under exclusive license to Springer Nature
Switzerland AG 2022
P. Prewett, *Foundation Mathematics for Science and Engineering Students*,
https://doi.org/10.1007/978-3-030-91963-4

Glossary

∃	there exists
∈	belongs to
∀	for all
$\mathbb{R}$	the Real Line
$\mathbb{C}$	the Complex Plane
(a, b)	Open domain on $\mathbb{R}$
$[a, b]$	closed domain
CF	Complementary Function
PI	Particular Integral
rhs	right hand side of an equation or inequality
lhs	left hand side
F	Farad (unit of capacitance)
Ω	Ohm (unit of resistance)
V	Volts (unit of potential)
A	Amp (unit of current)
e	e = 2.718281 . . . Euler's Number
e^x	(also written $\exp(x)$)
	exponential growth in x
$\lim_{x \to a}$	Value of $f(x)$ as x approaches infinitesimally close to a
ODE	Ordinary differential equation

Printed in the United States
by Baker & Taylor Publisher Services